MY SPY

MY SPY

MEMOIR OF A CIA WIFE

BINA CADY KIYONAGA

Perennial

An Imprint of HarperCollinsPublishers

NOTE TO READER: Some pseudonyms are used in the book
and are indicated with quotation marks at first mention.

A hardcover edition of this book was published in 2000 by Avon Books, Inc.

HarperCollins books may be purchased for educational, business, or sales promotional use. For information please write: Special Markets Department, HarperCollins Publishers Inc., 10 East 53rd Street, New York, NY 10022.

First Perennial edition published 2001.

Designed by Kellan Peck

The Library of Congress has catalogued the hardcover edition as follows:
Kiyonaga, Bina Cady.
My spy : memoir of a CIA wife / Bina Cady Kiyonaga.—1st ed.
p. cm.
ISBN 0-380-97587-4
1. Kiyonaga, Joseph Yoshio. 2. Intelligence agents—United States—Biography.
3. Kiyonaga, Bina Cady. I. Title.
UB271.U5 K57 2000
327.1273'092—dc21   99-050096

ISBN 0-380-79497-7 (pbk.)

01 02 03 04 05 JT/RRD 10 9 8 7 6 5 4 3 2 1

For Joe

I thank Theresa Park, my agent;
Jennifer Brehl and Jennifer Hershey, my editors;
and Rudy Maxa.

All have become my friends.

CONTENTS

*

"Bina, I want to stand up and be counted."

Joe spoke to me from his bed in Memorial Sloan-Kettering Cancer Center in New York City. My husband was dying of cancer. For twenty-eight years, I'd known very little about his work as a CIA operative. And now I was to hear his story—while there was still time. If there was still time.

He asked me to get a supply of legal pads and pencils. Dutifully, I would arrive at the start of visiting hours, at noon, and stay until eight at night. Joe dictated to the throb of the electronic monitor that pumped medication and nutrients into his heart.

The gray of the New York skyline, with its occasional flurries of snow, suited my mood; the quiet of the single room, with Joe's nurse stationed outside the door, ensured our privacy.

I listened as the pieces of our lives fell into place. Joe's frequent disappearances in Japan at the height of the Korean War; the midnight visits from General Manuel Noriega in Panama; the sudden appearance of submachine gun-armed guards around our home in Brazil—it was all beginning to make sense. I learned why Joe had encouraged some of my friendships, and discouraged others. And how something as mundane as my beauty makeover in Brazil had played a role in the overthrow of the country.

Windows that had long been slightly ajar were flung wide open; and as I wrote, I was, by turns, surprised, intrigued, appalled, delighted, and, sometimes, irritated.

≫⊷

Our story spans fifty years, four continents, three wars, a revolution, five kids, two races, and one faith. It is the story of an often stormy, sometimes blissful, but never dull marriage. It lasted thirty years—and then death did the parting.

Joe made me promise that I would see to it that our story was told.

After Joe died, he became the first CIA operative to be "surfaced"—his true occupation was revealed in his obituary in the *New York Times*. I was afraid then to publish my book—my children lived overseas and I was concerned about their safety and the safety of others with whom Joe had worked.

But the world has changed. And it's time to keep my promise.

Joe at the University of Michigan.

The United States did not deserve the old Central Intelligence Agency—the CIA. We, in the U.S., were too unsophisticated, too democratic, too open—and too dumb—to realize that for a secret agency to succeed it had to be just that, secret.

A lot of the criticism directed toward the Agency back then was well founded. Drug experiments, mob connections, and bungled assassination attempts were all sad mistakes, but they comprised such a small, bizarre segment of the Agency's work that in accentuating the bad, they obscured the good.

Looking back, I tend to think of the CIA in terms of crises averted rather than positive accomplishments, though there were plenty of the latter. It was the Agency's job to prevent the fire before it started, rendering a coup unnecessary.

In a business (the spy business) characterized by deceit, disinformation, and ruthlessness, the Agency tried to operate forthrightly and decently. It represented a country that was trusted and respected worldwide, and the CIA attracted like-minded souls.

We won't see their like again.

CHAPTER I

❧

BINA'S STORY

I WAS PRONOUNCED DEAD ON ARRIVAL.

While the hospital staff swarmed around my mother, I was baptized, placed on a gurney and ignored. I screamed my way back to life.

When my mother and father took me home to Harlem Avenue in Baltimore, I was still screaming. My mother had been trained as a nurse and approached baby care with nurselike dispatch. She later claimed I was no problem. She put me in the back bedroom, shut the door and let me scream. I screamed for four or five nights unceasingly, and then gave up. I could have given up the ghost—but I was no problem.

My problems surfaced later.

Never impudent or disobedient, I was simply contrary. Denied a wish, I would promptly throw myself on the floor, stiffen and stop breathing. Once my wish was granted, I would come out of it. Disturbed, my parents took me to see a Dr. Richards, chief psychiatrist at Johns Hopkins. I was three years old and still cute, with auburn ringlets and freckles. Dr. Richards suggested that she see me alone, since parents can be an inhibiting factor. It seems she lifted me onto her lap and began visiting with me very kindly. I kept my head down—I was having none of it. Finally, she addressed me directly.

"Now, Bina, I think you've been a naughty girl. You've made your mother and daddy very sad. What do you have to say to that? Bina, I'm speaking to you." (I remember. She was just too patronizing to suit me.)

I lifted my head, looked her full in the face and slapped her as hard as I could.

Later testing proved that I was normal, only too imaginative for my own good. It seems that I was dissatisfied with my parents, my surroundings, and, most especially, with my older sister and only sibling, Mary Ann. Once I'd come to terms with my environment, I'd come around.

Mine was a quiet, uneventful childhood—peaceful but dull. It didn't help that we were smack in the middle of the Depression and Prohibition. (Talk about a double whammy—first, you lose your job, then you can't even get a drink.) I remember a lot of potato soup at home, along with canned peas, canned pears, canned everything.

Fun for me was going with Daddy on Saturdays to the Lexington Market, a great Baltimore institution, for handcut potato chips and a horseradish-spiked crabcake sandwich, followed by a chocolate Coke at Ludwig's Drug Store. Or visiting the dank dungeons of Fort McHenry, home of "The Star-Spangled Banner," in Baltimore Harbor. The big night was Friday night—I'd watch my parents play "Russian Bank," a two-handed card game, while listening to *One Man's Family*, the radio soap opera.

I felt certain that there had to be excitement just around the corner, but there sure wasn't any on our block. So I took it upon myself to liven things up.

At three, I started with my catatonic fits. Mother was told to sprinkle cold water on my face, to no avail. Next, she was to *pour* cold water over my face—still no success. Finally, she was told to ignore me. That I couldn't stand. I snapped out of it.

And anyway, by age four, I'd discovered stuttering. I picked it up from Edward, the boy next door. Stuttering made everything he said seem so important. I think my parents were onto me by then. They patiently and silently forebore my halting efforts to get out even my own name: "B-b-b-bina C-c-c-cady." (I was really overdoing it.) Interestingly enough, it didn't hobble my singing at all. I could belt out a decent rendition of the World War I classic, "K-k-k-katy, Beautiful Katy." I'd learned it from my father, who had been jokingly serenaded with the tune by his Army buddies (being that his surname was Cady). Suddenly, one morning I was back to normal and never stuttered again.

Then the nightmares started. They were terrible. I can still see those little people as they'd troop across my bed. I'd squeeze them and they'd dissolve into blood in my hands. I found a sympathetic audience in my father. Daddy would sit on the edge of my bed, hold my hand and comfort me.

My nightmares stopped around age five. My parents were relieved—I'd finally "come to terms" with my environment.

Or maybe I was just biding my time.

✌

I WAS PRETTY MUCH OF A LONER, THOUGH I CAME TO ENJOY my solitude. (To this day I often talk to myself.) I was not close to Mary Ann, five years my senior. I always had the feeling she was ashamed of me. My strategy was to bribe her to keep me company. I would save my allowance and invite her to the Mickey Mouse show on Saturdays, and then she wouldn't even sit with me. She was everything I was not—short, cute, and popular. The only things we had in common were auburn hair and parents.

Mary Ann and I weren't just five years apart, we were light-years apart. We shared a bedroom, and when our grandmothers would visit (which was often), we shared a bed. At bedtime, Mary Ann would draw an imaginary line down the middle of the bed, and admonish me not to cross it. Occasionally, she'd wake me up during the night to let me know that I'd crossed it.

I suppose it was tough for Mary Ann to have a little sister

who always wanted to tag along. Despite that, we rarely argued—we weren't close enough. Much the same applied to Mother.

Thank God for Suzy, our cleaning lady. I would keep her company in the basement while she ironed, and help her pull the wash through the wringer. She would braid my hair into a mass of little pigtails. (I was way ahead of Bo Derek.) She braided my hair so tight that it hurt—but I thought I was the "cat's pajamas." The neighborhood kids branded me a "pickaninny." I didn't care. I was Suzy's pickaninny.

Most of my friends were imaginary ones. I'd write them little ethereal notes and then elaborately fold and bury the notes in the yard in wooden matchboxes (they were Daddy's, who chain-smoked Chesterfields). I kept a careful treasure map of where my six or seven missives were buried, figuring that I'd go back some-day and find that some angel had mysteriously added something to what I'd composed. (I really thought I was tuned into the beyond.) Once in a while, one would pop to the surface when Daddy cut the grass.

I was the son Daddy never had. We were very much alike in build, thought and sense of humor. Have you ever been in a movie and wondered why some people laugh at the wrong times? Daddy and I laughed at all the same things. I recall one Sunday outing when we went to a beautiful old inn outside Baltimore for dinner. Daddy drove smartly up to the portico of the inn—knock-ing down one of its wooden pillars in the process. Somehow, they allowed us to be seated for dinner, on the condition that Daddy first put the pillar back up. By that time, Mother and Mary Ann were acting like they didn't even know him. They stayed inside while I, his everlasting pal, went out front with him to offer encouragement. He took off his jacket and started hammering diligently—and quickly. (We were both in a rush to get in for dinner.) He finally stood back to survey the results of his labor: one upside-down pillar.

It was Daddy who took me every other Sunday and Thursday evening to the symphony (he held season tickets to the National Symphony Orchestra and the Philadelphia Orchestra) and to Saturday softball games down in the gully at the foot of Edgewood Street. But the chief honor he bestowed on me was that of "capper" of our home-brewed beer in our cellar. My reward was a draft of homemade sarsaparilla. I later graduated to a "cheese glass" full of beer. Daddy and I would sit side by side and enjoy my first version of a cocktail party. He'd take a sip and say:

"Boy, that's good."

I'd echo:

"Sure is good."

I admired my father. Self-made, he inspired me to make something of myself. When he was sixteen, he lied about his age so that he could fight with the mounted engineers in Mexico, who were chasing Pancho Villa (they never caught him).

Daddy went on to serve in World War I as an infantryman under Gen. "Black Jack" Pershing. After the war, he worked as a reporter for the *Baltimore City Record*, a business newspaper. He even took a Dale Carnegie course and swore by its effectiveness.

At 6'3", Daddy was a presence. And my protector. I remember that I was around eight when I decided I'd found religion. I started walking to daily Mass at 7:00 A.M. when it was still dark. Daddy, no daily communicant, would tag along. I assumed he'd found religion, too. Much later, it occurred to me that he had gone along to watch over me.

Daddy's pet name for me was "Sockalavitch," though I have no idea what it meant. He'd ride the Poplar Grove trolley home and when he'd round the corner at Rodbell's Grocery, he'd whistle and call out: "Heeeey, Sockalavitch!" I'd run down the alley to meet him, clutch his hand and proudly show him off to all the neighbors as we headed home.

Mother was kind, just not as much fun. She spent a lot of time "resting." She was closer to Mary Ann, so it was Mary Ann who got to study art and toe dancing. I was dying to study art, but was relegated to the piano (because of my long fingers) and tap dancing (my long legs, this time). And it was Mary Ann who got to go to Camp Pawatanika during the summers. I begged to go, and was finally sent. I was never so bored. Notwithstanding my Indian heritage, I discovered I had no interest in tepees or "powwows."

❧

I GUESS, THOUGH, THAT SOMETHING ABOUT CAMP MUST HAVE rubbed off on me. At any rate, I rounded up a few girls in the neighborhood and we officially formed the "Wig-Wam Club." There were five members. We were bent on good works and decided to give Christmas baskets to the poor. We canvassed the neighbors and were amazed at everyone's generosity, given the Depression economy. We gathered enough food for five Christmas dinners. Our mothers threw in toys and candy for the children. Our respective priests and ministers (the Wig Wams were multidenominational—one of our members was a Lutheran) supplied names of the poor families in the area.

I'll never forget that Christmas Eve when we delivered those baskets. It was kind of embarrassing to see children's eyes light up and mothers tear at the sight of their neighbors' bounty. I was especially embarrassed when one of the families turned out to be the Doyles. Mildred Doyle sat next to me in school. I felt bad for both of us when she ran from the room upon learning the reason for our visit.

Despite its success, the Wig Wams disbanded soon after that first Christmas.

❧

MOTHER AND DADDY COULD ONLY AFFORD WEEKLY FIFTEEN-minute piano lessons for me at Baltimore's Peabody Institute. For my practice sessions, they went to the trouble of renting an upright piano. Its location in our house could have been better. The piano seemed to glare at me in silent reproach from the corner of our dark, rather spooky cellar, irritated, I suppose, at the indignity of being placed next to Daddy's still and the laundry tubs. The bare bulb dangling overhead added to the macabre effect. I played to an audience of creatures lurking among the ceiling pipes. (Must have been those dreams again.) Practice was not my top priority.

I remember well Mother's admonition:

"Bina, if you don't spend more time at that piano, you're going to come home one day and it'll be gone."

I came home one day, and it was gone. The only piece I had managed to learn was Edward MacDowell's "To a Wild Rose."

I have to be fair: Mother tried. I remember that once, on my birthday, she went to the trouble of baking a miniature apple pie and hiding it from me. I searched for an hour and eventually found it in the RCA Victrola where the records should have been. For me, that was a grand event, better than if she'd had Shirley Temple tap-dancing up and down our back steps, and Deanna Durbin (whom I thought I looked and sounded like) singing to me from the middle of Harlem Avenue.

I felt for Mother. I think she was a little intimidated by the world. Some summers, we stayed at Sherwood Forest, a pleasant summer resort near Annapolis. Our cottage was near the gate-house, not on one of *the* seven hills, i.e. Robin Hood, Maid Marian, etc. Mother was so pleased when some of the ladies who really belonged invited her to join them at the clubhouse for bridge. After the game, I happened to be in the bathroom when the women came in and started talking about Mother. Apprehensive, I stayed in the stall and listened. They really panned her.

"Oh, she seems nice enough . . ."

"She's *nice* enough, but did you get a load of that purse?"

It went on for a while as I stood there, frozen, barely able to breathe. I was embarrassed for myself because I didn't have the courage to speak up, and embarrassed for Mother, even though she was a cut above those women. You see, she never would have spoken that way of someone. I never let on to Mother.

❧

MY MOTHER, ETHEL AGNES DOLAN, WAS BORN IN NOXON, Pennsylvania. She was one of thirteen children. Her parents,

Thomas G. and Sabina Rogan Dolan, moved their brood to Montrose, Pennsylvania and opened the Montrose Inn (no longer in the family's hands, sad to say). Each of the boys had a horse, and cut pretty dashing figures on their steeds. There were plenty of Airedales running around, bred by my grandfather.

Two of my mother's brothers, Edwin and Paul, had served in World War I. Both were gassed and held prisoner by the Germans. They had returned from the war broken men, but were still fun to be around, and funny. (I remember Uncle Eddie, after a long night on the couch, picking up his hairbrush one morning, holding it up like a mirror and commenting: "God, do I need a shave.")

It was during summers that I'd get to see my uncles and the rest of my extended family. We'd gather at my Uncle Timmy's farm in Dimock, Pennsylvania. There were Dolans, Rogans and Manleys everywhere you turned—they were all family, and all Irish. We'd get our mail from the postmistress, my cousin Lucille; our Sunday sermons from the parish priest, my cousin Donald.

It was a working farm, with cows, pigs, chickens, corn, fruit trees, and a vegetable garden. Everyone had chores; Mary Ann and I pitched in to help Nana Dolan and Aunt Aggie bake bread and churn butter. Everything was homegrown—the fresh eggs at breakfast, smoke-cured bacon, the cream for the oatmeal, and the canned peaches. The hash brown potatoes we didn't grow. We were Irish, but not that Irish.

Saturday nights were the payoff. We had to go outside the family for entertainment. His name was Jahael Kirkland, the blind fiddler, straight from County Cork. We'd push the table back in the kitchen and start tapping our feet to the jig. Anyone within shouting distance was welcome. It was a real honor to have Jahael Kirkland play. Uncle Timmy would even shave and put on a fresh shirt.

My name, Bina (as in "Dinah"), had been in my mother's family for generations—as far back as my great-great-grandmother. It was originally Sabina back in Ireland (after Saint Sabinus, a third-century martyr from Umbria who was beaten to death with clubs for confessing his faith). It was shortened to "Bina" when I was baptized, at Daddy's request. I think I like Sabina better.

My father, John Clayton Cady, was also from Montrose. His father, Alonzo B. Cady, had been a detective on the Pennsylvania State Police force. His pursuit of a member of the Black Hand took him to New York City. No sooner had he checked into his hotel and entered his room, than someone stationed outside his window shot him. He was dead before he reached the hospital. My father was twelve years old. His mother, Minnie Van Allen Cady, remarried, but later divorced and took back the surname Cady.

Both of my grandmothers were admirable women. Nana Dolan, recently of the landed gentry, was very impressive and very Irish. She was a handsome woman, white hair piled high on her head. She had been widowed young; many of her children had either died or proven to be big disappointments; she had fallen on hard times. Every setback served to stiffen her spine. She just held her head that much higher.

Nana Dolan's proudest possession was her Persian lamb coat which boasted a huge mink collar and cuffs, a gift from her daughter, Marguerite. How well I remember attending Sunday Mass with her. As she would approach the altar to receive Communion, everyone made way. She may have been down on her luck, but her position in the community remained unchallenged.

Grandmother Cady was equally good-looking, but hers were dark good looks. She was one-eighth Cherokee and looked it. She also wasn't too crazy about taking baths.

My mother was a great daughter-in-law. I remember the tactful way that Mother would suggest to Grandmother Cady that she change her housedress after it had been worn for a few days. I could have learned a valuable lesson from my mother, but I didn't.

Grandmother Cady would spend a couple of months with us each year. Her afternoons were spent casing every graveyard within walking distance to find a bench to commune with the long gone. That she knew not a soul, living or dead, in the city of Baltimore made no difference to her sport. It was left to me each dusk to round her up and bring her home. To give myself courage—and give her fair warning—I used to wear roller skates and blow a police whistle as I went from one burial ground to the next.

❧

IF MOTHER AND DADDY WERE WILDLY IN LOVE, THEY DIDN'T show it. But they were good to each other. I recall no upsets. Except one:

I was nine when Daddy worked at the Federal Land Bank. He was in the hallway one day visiting with the president of the bank, a close friend of his, when the president keeled over and died of a heart attack. Poor Daddy. The experience profoundly affected him, and he suffered a minor nervous breakdown. To get away, Mother, Daddy and I took off for our cottage at Sherwood Forest. Things were tense and Daddy'd had enough—especially of Mother. He finally said that he was leaving and stormed out of the cottage.

When he gunned up the car, Mother ran me outside and made me stand, alone, in the middle of the road in the path of our oncoming Model T. I remember so well the glare of the headlights, the sound of the speeding car—and the fact that I experienced no fear. I guess I figured that if Daddy was leaving I wasn't interested in hanging around either.

Daddy never tried to leave again.

❧

I CALL MYSELF AN IRISH CATHOLIC, ALTHOUGH MY FATHER was Welsh-Cherokee and a former Baptist. Mother asked that I never mention the Indian part of my background—now it's fashionable, but then it was an embarrassment. There's no question about my being a Catholic, though. I'm of the old school. I'm opposed to birth control, abortion, and in favor of the spirit of sacrifice— which our present Church lacks.

I attended Catholic schools, starting with Saint Edward's Grammar School, where I enjoyed the dubious distinction of being the only student to be dismissed from the church choir. Mine was a demoralizing influence, since I was given to sobbing at weddings and funerals and took most of my choirmates along with me.

The nuns really put the fear of God (and hell) into me. I recall committing only one noteworthy sin during my childhood. I wasn't sure if the sin was serious or not and decided to check with my mother. Her reaction was harsh. I was going straight to hell; there was no hope. So violent was her reaction that I sought to distract her. I rushed into the kitchen and stuffed tinfoil up my nose, both nostrils. If I was going to hell I'd do it with a flourish. I guess I thought the foil would stop my breathing. (I overlooked the fact that I'd always been a mouth-breather.) I kept on breathing but I started bleeding. The doctor was called. Order was restored, but I was in disgrace.

Saturday and confession time could not arrive soon enough to suit me. First, I told Father Fisher of my sin and was forgiven. Next Father Toolen. Then Father McShane.

Careful reflection told me that all three priests had accepted my confession too calmly. Perhaps they hadn't understood me. Perhaps it needed rewording. I mean, all bases had to be covered. I wasn't about to land in hell on a technicality.

I lined up to confess to Father Fisher once more.

No sooner had I told him of my sin than he interrupted me:

"Bina, is this the same sin you told me about earlier this evening?"

(How unsettling to be spoken to by name in the darkness of the confessional.)

I admitted to my duplicity.

"Bina, please go and sin no more."

Me, sin? It wasn't worth the effort.

TRINITY PREPARATORY SCHOOL IN ILCHESTER, MARYLAND was a big mistake for me. I only attended because I was awarded a half-scholarship. The nuns were cloistered and found me too outspoken for their tastes.

I remember one particularly painful episode during my senior year. I was president of the Student Council. Graduation was approaching (thank God!) and, as was their wont, the nuns made all of the decisions concerning the event itself. We, the graduates, were appalled. We were not consulted, even as to what we'd wear. The nuns had chosen an outfit that was white, uniform and ugly. They were trying to turn us into a bunch of mini-nuns.

I was furious and told the headmistress so. My fury paled next to hers. A formidable woman when happy, Mother Frances that spring afternoon was downright scary. I have to hand it to her, though. She was in perfect control as she came to the point:

"The graduation gown has been decided upon and you are to accept that decision."

Mother Frances's tone and expression were unmistakable.

Undeterred, I replied: "Mother, what's the percentage of having a Student Council if we have no voice?"

That did it. She'd had it with me.

"Young lady, you've spoken out of turn once too often."

The next thing I knew I was attending a specially called Student Council meeting. Present were all the faculty and three Student Council members. I guess the others weren't invited—the ones who were my friends. Mother Frances called the meeting to order, and, in a flash, I was voted out as head of the Student Council. I received only one vote—my own. Why they didn't throw me out of the school as well, I'll never know. They would have done us all a favor.

The only really good time I had that year was when I met Bill Linde at a USO dance. It was 1942; I was sixteen years old. "Linde" was a cadet with the Merchant Marine. I liked him a lot. He was tall, cute, fun to be with—and thoroughly decent.

We talked and talked. He commented once, out of the blue, how unbearable it would be to have your child die before you did. It was an unusually thoughtful thing to say, and made an impression on me.

Within weeks, he was shipped out on the treacherous Murmansk run across the North Atlantic. A few months later, I received a letter from his mother. Linde's ship had been torpedoed by a German U-boat and sunk. On the life raft, he had volunteered to stand and hold the lantern aloft so that they could be spotted by a rescue ship, while his shipmates huddled together. He had frozen to death, and a swell had washed his body overboard. He was twenty-two.

Later, I was invited to attend his memorial service at the Merchant Marine Academy on Long Island, where he was decorated for heroism. His mother gave me a silver-plated tea service he'd won in some athletic event. It's in my dining room today.

※

MOUNT SAINT AGNES JUNIOR COLLEGE WAS MORE TO MY liking. I was a boarder, but despite that, I felt as though I'd been released from prison. The Mercy nuns treated us as adult women and we didn't betray their trust. While at The Mount, I got really big news: Daddy, who by that time had joined the Foreign Service, was to be stationed as Agricultural Attaché in Bogotá, Colombia. I was so proud of him: my father, who'd only had two years of high school, was to have two Ph.D.'s serving under him. The excitement—for him and for me—wasn't around the corner, it was halfway down the world.

Mother's initial reaction was less enthusiastic. Harlem Avenue may not have thrilled her, but at least it was home. Learning Spanish, making new acquaintances, starting all over again—these things were a challenge. I watched Daddy as he listened to her concerns and could almost see his vision of the future dim.

But I had underestimated Mother. She rose, hesitantly, to the occasion. She even changed her hairstyle. I remember going with her to Schwartz Brothers (now Joseph Banks Clothiers), who put together a whole ensemble: a black gabardine suit with matching flowered toque. My mother was venturing out into the world; I only wish she hadn't waited so long. When she and Daddy turned around and waved to me as they boarded the plane, they looked for the first time like a real couple.

This was the summer of 1945, between V-E and V-J Days. Mary Ann had joined the WAVES and had married a Navy doctor, Frank Cyman of Detroit. By now, Mary Ann and I had become friends; actually, close friends. We no longer shared a bed and bedroom. In our case, distance did make the heart grow fonder.

I was to join Mother and Daddy in Bogotá later that summer. Mother kindly suggested that I bring a friend along, so I invited Rita, Mary Ann's sister-in-law, a happy choice. We were about to be set loose in the High Andes of Bogotá, an Old World Shangri-la of majestic bullrings and rococo cathedrals.

Mary Ann's parting words at the airport were that, should she become pregnant, she would send a cable "congratulating the grandparents," considering it indelicate to flatly state that she was pregnant. In the excitement of our good-byes I forgot the code.

Even the Pan American flight down was memorable. The pilot, George Cauthen, a farm boy from North Carolina, invited me up to the cockpit as we neared Bogotá. After a little while, he offered up the controls of the plane. Talking and laughing, I maneuvered our unsuspecting passengers over the green, jagged mountains below. Someone caught on—I suspect the copilot—and Pan Am grounded George for a week.

Rita and I were a novelty (she a blond, me with my red hair) among the raven-haired beauties of Bogotá. And I soon met up again with George who, this time, let me drive his black Packard convertible. I liked George—but I *loved* that car. We never were formally engaged, but were more or less committed. At least I was.

In due time, a cable from Mary Ann arrived in Bogotá "congratulating the grandparents." Mother and Daddy threw a big party that night but couldn't figure out why Mary Ann had failed to state the sex of the child, and happily overlooked the fact that she'd only been married five months.

❧

I HAD INTENDED TO ATTEND WELLESLEY FOR MY JUNIOR YEAR but decided on the University of Michigan instead. Mary Ann was lonely in Detroit. I was all set to start in the fall, until a raw oyster intervened.

There I stood in a wonderful purple wool-jersey draped gown, scooped neck, long slender sleeves, cinched with a wide leopard belt. My final fitting—the gown was for the last ball of the summer. I felt woozy and told Mother so. I wanted to sit down and rest. But Mother told me it was all in my head. Her words were the last thing I heard before collapsing in a heap.

My next memory was of being home, in my own bed, surrounded by a cluster of concerned faces. Tests later determined that I had contracted hepatitis, an especially virulent form, from eating a raw oyster in the inland heights of Bogotá.

Bed rest and massive doses of vitamins were prescribed. No problem—getting out of bed was impossible. Imbibing vitamins with orange juice and raw eggs was more of a problem. I couldn't keep them down.

Weeks—no months—later, Mother and Daddy (and George) decided that the hepatitis had run its course. The doctors weren't quite as convinced. Truth was, I was driving them crazy and they knew if I didn't leave soon, I'd be too late for the spring semester at Michigan. I had already missed the fall semester.

But my yellow skin and even yellower eyeballs would have to get past Immigration. Mother came up with the solution. She dolled me up in a huge head scarf that covered most of my face and sunglasses the size of saucers.

I made it through Immigration and landed on my wobbly feet at Ann Arbor. It was snowing and I was all alone—ready for my next adventure.

His name was Joe.

❧

CHAPTER II

JOE'S STORY

JOSEPH YOSHIO KIYONAGA WAS A REMARKABLE MAN. HIS WAS a magnetic, complex personality. I never completely figured him out.

When I met Joe in 1946, he stood 6'4", spoke perfect English, had fought in Europe and was studying law.

He was urbane, tremendously attractive—and just plain sexy. I was fascinated.

But the Joe Kiyonaga that I met had come a long way from home.

❧

IN 1866, HIS ROYAL HIGHNESS KING KAMEHAMEHA OF HAWAII decreed that a colony for lepers be established. Those ravaged by the dreaded disease were to be hunted down, captured, and banished to the colony, never to return.

The site had to be remote, virtually inaccessible, allowing no escape. Chosen was the barren, flat Kalawao peninsula jutting out from the Hawaiian island of Molokai, bounded on three sides by shark-infested, wind-thrashed waves that lumbered south from the Aleutians. Abutting the peninsula were the world's tallest sea cliffs, towering over 3,000 feet—impossible to scale unless you happened to be a goat.

The rest of the island, sparsely populated by pineapple-field laborers, the lepers called "Topside." Joe just called it home.

⚓

JOE WAS BORN ON OCTOBER 31, 1917. (HALLOWEEN—A pretty apt birthday for a future spook.) An only child, his parents were Japanese immigrants who barely tolerated each other.

Joe's mother, Waki Yamaki, had set sail from Japan with her parents and seven brothers and sisters in the early 1900s. Their destination: Los Angeles. This was not a pleasure cruise. Economic times in their home city of Hiroshima were rough. Relatives already in California had sent back glowing accounts of fortunes to be made.

The ship's quarters must have been cramped. At the stopover in Honolulu, Waki and her brother were dropped off (okay, dumped) in the care of friends "to receive an education." Imagine watching *that* ship steam away from port. Waki was eight.

Joe's father, Junzo Kiyonaga, didn't have much more of a promising start. His family back in Japan wasn't doing any better than Waki's. So at fifteen, Junzo was sent to Maui, Hawaii, to work in the pineapple fields as a contract laborer. The terms of the contract were simple enough: ten years at an almost nonexistent wage. (Talk about the American Dream.) Room and board, of sorts, were to be provided. The work itself was brutally unrelenting—hours slashing sinewy pineapple stalks under a blazing sun.

Junzo had a choice at the end of the ten years: he could return to Japan or remain in Hawaii. Junzo chose to stay in the islands.

The Kiyonaga name is an illustrious one in Japan. A samurai clan, "Kiyonaga" means "everlastingly pure." The significance of the name must have been lost on one of Junzo's ancestors, Torii

Kiyonaga, an eighteenth-century woodblock print artist and one of the five great ukiyo-e artists. Kiyonaga depicted women. His works bordered on the pornographic and appealed to the baser elements of Japanese feudal society.

Waki had received only three years of formal education in Japan. Once in Hawaii, her education stopped. After her parents left, she was pressed into domestic service at age nine. Resourceful, she studied at night. She taught herself English and the American Palmer method (of handwriting), and also mastered the art of "kanji" (Japanese calligraphy).

Waki eventually came to work as a cook at the leper colony on neighboring Molokai. The place at that time was run by a Dr. Goodhue, a distant relation of Ralph Waldo Emerson, who made great strides curing leprosy through surgery. Molokai lore has it that he was one of the few doctors who didn't treat the lepers like lepers.

Goodhue was in the tradition of Father Damien, who had come to the colony in the 1870s to dedicate his life to the lepers. Father Damien eventually contracted the disease, revealing his condition to his congregation one Sunday when he began his sermon, "My fellow lepers . . ." In the 1920s, Goodhue himself would end up leaving the colony, bound on a freighter for Shanghai. It was suspected that he had contracted leprosy.

Waki was introduced by a mutual friend to Junzo, who, by that time, had finished his labor contract. They married, and, initially, settled on Maui.

After Joe was born, the Kiyonaga family moved to Molokai. As virgin territory, the island appealed to Joe's father. Before long, Junzo had established himself as a resident car mechanic and introduced the first taxi service (one car!). Waki became a seamstress and opened Molokai's "premier" dress shop, which consisted mainly of muumuus, aloha shirts and souvenir pillows. Over the next fifty years she'd make virtually every "holaku," or prom dress, on the island.

For extra income, they ran a bathhouse where the exhausted laborers could drop by and relax in a steaming cauldron, Japanese-style, and enjoy a potent cup of sake produced by a ramshackle still. Junzo handled the still detail.

The Kiyonagas were poor. Home was a shack on stilts with poultry and swine roaming the underpinnings. No electricity, no plumbing. Clothes were homemade—and scarce. Joe went barefoot until he finished grammar school. He received his first pair of shoes as a graduation gift. They were a $3.99 Sears special, size 9DDD, just to be on the safe side.

❧

WHEN JOE AND I MET, NEITHER OF US WAS ANXIOUS TO REGALE the other with stories of our childhood. Joe told me only of one memory: Saturday nights, when the Kiyonaga cabin was transformed into the "Paramount Theater" of Molokai.

Junzo had bought a movie projector (his enterprising spirit again) and arranged with a distributor in Honolulu to send him used films. Lacking a screen, Junzo inventively strung up a bedsheet along the side of their house. The crowd would assemble in a sandy clearing amidst acacia trees and flickering kerosene torches.

There was something for young and old. Ice cream (handcranked) and corn whiskey (courtesy of the Kiyonaga still) were served, for a price. Joe's best memories were of barely discernible stagecoach Westerns illuminated by a full Hawaiian moon, the gunfire and clopping of horses' hooves often eclipsed by the pounding surf. Right then, Joe's life-pattern was set: the good guys, the bad guys, the hero and the heroine—and, always, the happy ending.

❧

I HAD TO SMILE WHEN JOE TOLD ME ABOUT MAKING THE ICE cream for the Saturday night movies. Despite our vastly different backgrounds, our Saturdays were pretty similar.

He was cranking out ice cream while I was staging a two-penny show for the neighborhood kids in our Harlem Avenue basement—

while each of our fathers was busy at his respective still. With my
hard-earned bounty I'd buy a "snowball" —shaved ice with choco-
late sauce and marshmallow topping—and head for the show at the
Astor Theater. I may have even seen a Western.

☙

FOR YEARS, THAT WAS ABOUT ALL I KNEW OF JOE'S CHILD-
hood. I almost wish it had stayed that way. Only later did I learn
the truth.

While I was sitting around doing nothing in Baltimore, Joe
was living out his private hell of a childhood. It would start each
morning, when Joe would watch his father have his breakfast in
bed, straight from the bottle he kept within fumbling reach. Back
from work, Junzo would return to his bed and to the bottle.
That's when it usually began: the slurred shouts, the fights, the
blows inflicted on Waki. Joe, a brave boy, would try to get between them—he was all she had. Once Joe even took a picture
of Junzo sagging with drink, and scrawled across the bottom
"Why Dad?"

But Joe already knew the reason. He was the reason. Although
Joe looked like a little Buddha when he was born, Junzo had
watched him grow tall . . . very tall. And he was different looking
from the other Japanese kids. It's easy to condemn Junzo, but
that man suffered.

Joe must have sensed it—he even looked a little uncomfortable
in his third-grade class portrait at Kamalo Elementary. As the
University of Hawaii newspaper later remarked concerning his
performance as Coriolanus in the annual school play: "Kiyonaga
was noticeable."

People gossiped. Molokai combined the paranoia of an insular
mentality with the lassitude of a sleepy Texas town. Everybody's
life was everybody's business—and business was bad. Rumors
flourished, and Joe was a rumor incarnate. At eight years old, he
took to carrying a gun to defend his mother's honor.

❧

I WAS NEVER REALLY SURE JOE WAS ALL JAPANESE. I DON'T think that he was all that certain, either. But poor Joe. He'd taken up the cudgel to defend his mother's honor early on and never put the damn thing down. We spent half of our married life explaining Joe's Japaneseness to people and the other half fighting about his mother. In between, Joe managed to become an unqualified success as a CIA operative. Here he was gathering information on the intimate details of people's lives, yet the critical details of his own background always eluded him.

It's entirely possible that Joe was all Japanese. His mother and father could have suffered needlessly—and Joe, too, for the matter. But explaining Joe's being 100 percent Japanese got to be quite an undertaking. People never failed to question Joe about his background once they heard his name. Joe would simply state that his mother and father were Japanese, period (looking slightly annoyed). Not me. I had to go him one better. His family, a samurai clan, had traditionally been part of the Emperor's Imperial Guard. You know the Imperial Guard? Well, they're an elite group, all more than six feet tall—quite a feat in ancient Japan—and Joe's family for generations had helped man said guard.

Then there was Joe's uncle, Joe's mother's brother, who visited us when we were stationed in Japan. A handsome man, he stood 5'10". In the retelling, I had him assume the proportions of a sumo wrestler—if not in girth, at least in height.

So busy was I in covering Joe's tracks that I finally placed his origins in Hokkaido, northernmost Japan—the land of the Ainus—taller-than-usual Japanese and often red-headed. In actuality, of course, Joe's mother was from Hiroshima, her brother a member of the city council. Joe's father came from Fukuoka in the south of Japan.

Once committed to my task I simply could not give up.

Dear friends and passing acquaintances (God knows, anyone who would listen heard my tale), please forgive me. I don't even know if there was an Imperial Guard. If one existed, I feel certain the Emperor would have wanted them to be tall and handsome and of samurai stock. But the Kiyonagas were never among them.

≫⊷

GIVEN THE CIRCUMSTANCES, JOE MADE FEW FRIENDS. THE exception was Mike McCorrestor, the closest neighbor, a mile down the road. Joe also befriended a longtime boarder his parents had taken in, an older Eurasian man. It was Joe who found him one morning, slumped over his steering wheel, the carbon monoxide seeping in from the hose connected to the exhaust pipe. I did not learn of this event until recently, from one of my sons. I find this fact alone significant. Why didn't Joe tell me?

Books were Joe's refuge: biographies of Lincoln, accounts of the Prussian wars, Sherlock Holmes mysteries and Zane Grey's *Riders of the Purple Sage*. Exceptional men seem to have that trait in common—as children, they devoured the public library.

Hunting was his other escape. This wasn't a hobby. It put food on the table. On his way home from the pineapple fields (his summer job was to weigh the pineapples), he'd use an ancient .22 to shoot morning doves or Franklin partridge from the bed of the truck. This was the Molokai he loved.

Other times, he'd venture into the hills in search of a more elusive prey: mountain goat.

≫⊷

IT WAS AT DAWN, AS JOE DESCENDED MOUNT KAMAKOU, the highest peak on Molokai. He'd spent the night there after a day hunting for goat. He was making his way down the trail, eager to show his parents his first kill. It was tough going, as his trophy weighed as much as he did. He spied a familiar figure approaching him—his father, bearing a small lunch for them to share: nogiri (rice balls wrapped in seaweed), dried fish, green tea. They ate and talked before heading home.

It was one of two conversations Joe ever remembered having with his father. Joe was eleven.

❧

WHAT MUST HAVE MADE IT WORSE FOR JOE IS THAT MOLOKAI is so stunning. It's one of those places that seems to have no reason to exist, except to be beautiful. I can see Joe, at low tide, spearfishing alone among the lava rocks, oblivious to the torrid sunset behind him. And Molokai was so unspoiled—leprosy, and its attendant death, didn't exactly pack in the tourists. To this day, Molokai remains virtually undiscovered. It should have been paradise for a young boy.

Speculation could have it that some combination of lucky breaks, or even a special star, must have contributed to the creation of the Joe Kiyonaga I came to know. I don't believe it. I'm convinced that by sheer weight of determination, Joe made himself into the man that he became. He could have modeled himself after Thomas Jefferson or possibly Machiavelli (Joe always aimed high). He had what some might call a quiet reserve; I call it self-reliance.

It started at thirteen when Joe won an essay contest on automobile safety that snared him a scholarship to attend Lahainaluna, the "oldest high school west of the Rockies."

The school, still in existence, is on Maui. So primitive was transportation in those days that Joe traveled each year by sampan—a six-hour trip over rough seas—to his first day of boarding school. Although the travel was primitive, the school was not. H. Alton Rogers, the principal, was Harvard-educated and attracted a reputable faculty.

Joe took one look around, recognized his shortcomings and lack of polish, and set about correcting them. He approached the home economics teacher and suggested that she teach him and some of his like-minded friends the correct use of the knife and fork, as opposed to chopsticks, as well as proper table manners.

She kindly initiated a series of lunches during which the boys learned the niceties, at the home economics students' expense.

All boarding students at Lahainaluna had to work for their keep. Joe managed to secure the job of driver and yardboy for the principal and his wife. It was Mrs. Rogers who undertook to tutor Joe in speech. Until then, he had spoken a mixture of pidgin English and Japanese. Thanks to her, Joe traded in his pidgin English for a Brahmin accent.

His junior year, Joe faced the prospect of the prom. He didn't know how to dance, and neither did most of his classmates. As class president, he asked his English teacher to give them dancing lessons. She agreed and had the carpentry shop make 110 broomsticks. Every Tuesday evening a Victrola was set up in the gym while she led all 110 boys through the motions of the current dance steps—fox trot, box step, waltz—with broomsticks for partners.

⌇

JOE WAS A GREAT DANCER. TO SEE US ON THE DANCE FLOOR was a dead giveaway. When Mary, our oldest daughter, was considering marriage, she came to visit us in Panama to press her suitor's claim. He was tall, attractive, a Princeton graduate, a law student, a member of the law review, and recently apprenticed to a Federal judge. Joe smiled reflectively, sipped his cognac, took a puff of his Montecristo (a recent gift from General Omar Torrijos), and asked,

"Yes, Mary, but can he dance?"

⌇

As Joe's GRADUATION APPROACHED, YET ANOTHER HURDLE appeared: the traditional dress for graduation was white jacket and navy pants. Joe owned one suit—a navy jacket and white pants—handmade by his mother. He knew just what to do. As student body president of the Class of 1936, Joe introduced a new official dress code: navy jacket and white pants. (We were destined to be together; how many people do you know who had so many problems with their graduation outfits?)

On the big day, Joe's parents made their first trip to Lahaina-luna. They watched as Joe collected award after award. They must have realized that they were beginning to lose him.

The University of Hawaii is beautifully situated in Manoa Valley on the island of Oahu. Joe wanted to study medicine, but his mother preferred that he become a teacher. He acceded to her wishes, working his way through the University of Hawaii Teachers' College as a houseboy. He also undertook the job of rent collector in a substandard neighborhood.

❧

YEARS LATER, HIS STINT AS A RENT COLLECTOR WOULD come back to haunt him.

Apparently, one of the tenants was uncooperative—actually a group of girls. After several unsuccessful attempts to collect overdue rent, Joe made one last bid. No one was at home, so Joe simply confiscated the clothing he found in the closets as security. He was struck by their finery and only upon reflection realized that he'd robbed a brothel. He was called into court and received a suspended sentence.

His brush with the law cropped up later when he applied for work with the Agency. Joe neglected to report it; the CIA investigator unearthed the incident and questioned him about it. Joe said he believed he'd committed no crime. He considered it not worthy of mention.

❧

IT WAS AT THE UNIVERSITY OF HAWAII THAT JOE FELL IN LOVE for the first time. According to Joe, Lei was a dusky, lithe Hawaiian maiden. Bear that picture in mind. I know I did.

He also fell in love with golf. He played each weekend, a break from his teaching job. He was playing an early round one

Sunday morning when he realized his life would never be the same. It was December 7, 1941.

Five thousand miles away, I was listening to a symphonic program on the radio when the music was interrupted with the bulletin: the Japanese had attacked Pearl Harbor. The next day, I, like millions of other Americans, strained to hear FDR's words over the radio as he proclaimed the "date which will live in infamy."

Joe couldn't believe it. He and other Japanese-Americans reeled from the atrocity of the attack, especially the sight of black smoke billowing from the sinking USS *Arizona*. Thousands of Americans had been killed and their home islands of Hawaii invaded. Worse yet, the invaders shared, with them, a common heritage.

Meanwhile, the country, very naturally, was whipped into an anti-Japanese frenzy. The Nisei (Japanese-Americans) were immediately suspect. Martial law was declared on Hawaii, and the Hawaiian Nisei, including Joe and his family, were shadowed by the FBI. Respected intellectuals like Walter Lippmann advocated evacuation of Japanese-Americans from the West Coast, which was considered vulnerable to attack. The media led the chorus. This quote by Henry McLemore of the San Francisco Examiner was typical: "I am for the removal of every Japanese on the West Coast to deep into the interior. Herd 'em up, pack 'em off, and give 'em the inside room in the Badlands. Let 'em be pinched, hurt, hungry, and dead up against it." It didn't seem to make any difference that these were Japanese-Americans.

It did to Joe and his friends. Born dual U.S.-Japanese citizens, they had elected to become American citizens on their twenty-first birthdays. They were ready to fight—to defend their country and prove their loyalty. The Nisei petitioned President Roosevelt to allow them to form their own combat team. On January 22, 1943, the 442nd Regimental Combat Team was born.

World War II was the best thing that ever happened to Joe; it got him out of Hawaii. The first contingent of Hawaiian Nisei, Joe among them, left for training at Camp Shelby, Mississippi, in early 1943.

The experience of traveling was traumatic for the boys from Hawaii. Joe and his friends all came from tiny islands surrounded by the vast Pacific. Just to arrive on the Mainland took five days by ship—the *Lurline*. Then began the train ride to Mississippi, eight days and nights in all. The psychological impact of going to bed at night on land and waking up in the morning, still on land, was tremendous. (It takes only about half an hour to drive across Molokai.) Their vision of the United States took on gigantic proportions.

Their sense of place went through other changes. It was in the Deep South that Joe first learned where he fell on the racial divide.

❧

ONE SATURDAY NIGHT, JOE AND A FEW OTHER NISEI WAN-*dered into a bar where some white soldiers and a smaller group of black soldiers were brawling. The Nisei jumped in on the side of the black soldiers.*

More than a year later, Joe, now an officer in France, was driving along in a Jeep and obliged a black soldier, hitching a ride. The soldier looked at Joe:

"Say, weren't you in Camp Shelby?"

"I was, yes."

"You jumped into a fight with a bunch of white soldiers in a bar. You remember?"

Joe thought for a moment:

"That's right—I remember."

His passenger nodded knowingly and replied:

"Yeah, us colored folks, we gotta stick together."

❧

TO HEAR JOE TELL IT, THE HIGH POINT OF HIS STATESIDE service occurred during the furlough he spent in Washington, D.C. He did the usual sightseeing, but that receded in importance next to a date that he had with a SPAR (the U.S. Coast Guard Women's Reserve). Joe had been billeted in the Hay-Adams Hotel and discovered that the entire second floor was occupied by SPARs. Never one to pass up an opportunity, Joe zeroed in on Karen, an attractive blonde. She agreed to have dinner with him that evening.

They had just been seated at a table at the Lotus Club, a real tourist trap, when the waiter appeared.

"Your order, sir?"

"Karen, would you care for a drink?"

"Thank you, Joe. I'll have a champagne cocktail."

"Make mine a scotch and soda."

Joe was in full command of the situation and relishing the prospect of the evening. Conversation was a little difficult with the band blaring in the background.

Joe shouted, "Karen, you have to be the best-looking sailor I've ever seen."

"I was selected 'Miss North Carolina' last year," was her modest reply.

Joe was really warming to his subject when the drinks arrived. Understandably excited, he downed his drink in a few gulps. Karen seemed to have eyes only for him. Or was it his glass?

"Joe, I've never seen anyone drink a scotch and soda that way."

"How's that?"

"Do you always drink the soda first?"

Only then did Joe notice that what he had been gulping was an iced glass of soda. The pony of scotch remained untouched.

Unfazed, Joe downed his scotch in one swig. He was never one to be outmaneuvered.

"Oh, yes, we drink them that way in Hawaii."

For the rest of the evening Joe was doomed to drinking "Hawaiian style." Karen was fascinated.

Joe claimed that he spent the rest of his leave avoiding her. (I question that.)

≫

THE "FOUR-FOUR-TWO" WAS STRICTLY A VOLUNTEER OUTFIT. It consisted of three infantry battalions under the command of Col. William Pence. (Senators Dan Inouye and Spark Matsunaga were among its notable members.) The enlisted men were Nisei; the officers were Caucasian, and clearly not volunteers. Joe was a member of the 3rd Battalion, M Company, and he commanded a heavy weapons section.

Soon the Hawaiian Nisei were joined by their Mainland counterparts. Despite being interned in prison camps, the West Coast Nisei also wanted to fight for their country.

A friendly rivalry had always existed between the two Nisei groups. The West Coast Nisei were called "Katunks," supposedly because when they found themselves in brawls and were hit on the skull, their heads went "katunk."

The Hawaiians were nicknamed "Buddha Heads," reputedly because they affected the shaven-head style of the Buddhist monk. Theirs was a practical consideration. Hawaii is warm.

The 442nd faced the Germans in the European theater. But Joe's first contact with the enemy had been at Camp Shelby where he had guarded German POWs. Joe got to know some of the prisoners. These were elite troops, all members of the Afrika Korps who had fought under Field Marshal Rommel in North Africa. They would labor all day in the peanut fields under the relentless Mississippi sun. Joe was impressed when, after nightfall on Christmas Eve, these same Afrika Korps members gathered and sang Christmas carols a cappella. The prisoners presented Joe with a valpack as a gift, his initials meticulously stenciled on it.

They may have been the enemy, but Joe couldn't help but admire them. He found them to be diligent and disciplined, even cheerful, under the circumstances. The thought of possibly killing their brethren in the coming combat saddened him.

The 442nd arrived in Naples on June 24, 1944, and was immediately thrown into the Rome-Arno campaign as part of the Fifth Army. It was the 442nd's baptism by mortar fire. They

proved formidable and fearless. There are accounts of German
soldiers watching in amazement as Japanese-American troops—
not looking like your typical kids from Kansas—fought their way
toward them. They helped push the Germans north to the outskirts of Florence.

I guess no amount of training can prepare a young man for
Florence. Or for Fiorella. Her father, a surgeon, had left Florence
with the rest of the family for the relative safety of North Africa.
Joe and Fiorella enjoyed the run of the family villa that over-
looked the city. I later replaced Fiorella in Joe's affections, but
Florence always remained his favorite city.

The regiment's second campaign was fought in the cold, heavily wooded and extremely rugged Vosges Mountains of France. A
stalemate developed; Allied troops were unable to dislodge the
German defenders. Two battalions of the 442nd converged on
Bruyères (the German headquarters) at dawn of October 18,
1944. The remaining battalion attacked the high ground north of
town. The conquest of the town, and the hills around it, took
three days.

During this campaign, the 442nd distinguished itself on a special mission. The "Lost Battalion" of the 141st Infantry from Texas
had been cut off and surrounded by the Germans. There was a
call for volunteers from the 442nd—and every man stepped forward. It took them eight days to rescue the Texans. The casualties
were staggering: one company in the 3rd Battalion was left with
seven men, another seventeen.

The morning following the rescue, Col. Pence called a forma-
tion. Once the troops had mustered, the colonel turned to the
regimental commander:

"I wanted the whole regiment here."

The commander started to cry.

"This *is* the whole regiment."

Every member of the 442nd was designated an "Honorary
Texan."

FIFTY YEARS LATER, MY SON PAUL WAS AT A DINNER IN Houston with a partner at a major Texas law firm celebrating a courtroom victory. Paul happened to mention that his father had served with the 442nd. The partner quelled the dinner conversation and proposed a toast to Joe and the 442nd. They don't forget in Texas.

※

JOE'S OVERALL REACTION TO HIS WAR EXPERIENCE WAS THAT he was constantly frightened and tired but proud of having served in the infantry: "The foot soldier sees the war from the ground up, but we all envied the Navy—they at least had clean sheets."

He told of going for twenty-eight days running without a bath. (The Japanese in him must have rebelled.) Sleeping was also a problem. Always one who enjoyed creature comforts, Joe invariably spent much of the night digging and preparing his foxhole. Risking life and limb, he would forage for hay from neighboring farms to feather his foxhole. (He was also his unit's prize chicken filcher.) War may be hell, but Joe made it sound more like a second-rate hotel.

He became his battalion's "Kilroy" —his buddies always knew when he'd preceded them. One soldier would notice a fastidiously designed—and long—foxhole and call to another: "Hey, brudda, looka dat. Kiyonaga was here!"

Joe received a battle field commission and was decorated for valor. To celebrate, Joe was invited to drink with some of his fellow officers, most of them Caucasians. Wood alcohol and juice didn't agree with Joe. Daybreak found him crawling along a dried river bottom, dodging bullets, trying to rejoin his company. The sight of their new commander making such an inglorious entrance into battle must have given his troops pause.

I know Joe took lives, but he never told me how he felt about that. It wasn't something you talked about. I do know that the first time he was under fire, he quickly scrambled to lay down a mortar barrage. He had to fire over a hill by dead reckoning. When 442nd troops charged back down the hill, asking who'd fired the mortars, Joe was fearful that he'd killed some of his own men. It turned out that he'd wiped out a German machine-gun nest.

Seems as though Joe excelled in combat at a distance; less so face-to-face. Joe spied a lone German soldier once on a ridgeline, an easy shot. He let him go.

Who knows? It could have been that same soldier that Joe ran into later. He was in a town, unarmed, looking for food, when he rounded a corner and almost ran into a German officer, a Luger pistol holstered at his side. This was it—Joe expected to die on the spot. The German reached for the Luger, handed it to Joe, and raised his arms over his head in surrender. Joe took him prisoner. (My son, John, now has the Luger.)

I never would have known Joe Kiyonaga if he hadn't dived behind a boulder at the precise right second—one of his buddies, right next to him, died from a wound to the head. Another time, Joe took a fortuitous trip to the latrine in the middle of the night and returned to find his foxhole demolished by a German mortar. His survival of the war seemed so random. Then again, so did everyone's.

The 442nd had chosen for its motto "Go for Broke." It did. They emerged from the war the most decorated and decimated unit of its size in U.S. military history.

On the eve of V-E Day, Joe received a telegram from home notifying him that Junzo had died.

❧

THE NIGHT BEFORE JOE HAD LEFT FOR THE WAR, JUNZO HAD walked him to the end of the pier to await the boat that would take him on to Honolulu.

Junzo told Joe that he loved him and was proud of the man Joe had become. He asked Joe to take care of his mother, upon his return.

Perhaps he had a premonition. They talked quietly until it was time for Joe to go.

As Joe boarded the boat, he knew he would never again see his father. And that Molokai would never again be his home.

﹋

CHAPTER III

THE LIGHTS GO ON

WHAT BETTER PLACE TO FALL IN LOVE THAN ANN ARBOR, Michigan? The University of Michigan campus in the fall of 1946 was alive with color. It was also alive with returned veterans. They outnumbered the girls twenty to one. I arrived there practically engaged to George, the Pan Am pilot, but was doubly blessed. I was housed in the Martha Cook dormitory, the honor dorm for women and sister building to the neighboring Law Quadrangle.

The day I arrived for registration, I found myself in an enormous auditorium filled with thousands of students discussing course availability, majors and professors. This was serious business. Somewhat overwhelmed, I approached a counselor and asked what the easiest course of study might be.

He took one look at me.

"Drama."

And that was that.

Every fall the Law Quad and Martha Cook held a traditional mixer-dinner. The couples were prematched, not by computer or questionnaires, but by common sense and an astute pair of committees. The couples were so well paired that, reputedly, an average of thirty-two engagements resulted yearly. I didn't sign up for the dinner because I wasn't dating. The dinner was scheduled for Thursday, November 6.

Monday evening, three days before the dinner, I was comfortably settled in my room reading a book (I still wasn't into studying) when there was a knock at the door. I found myself confronting Margie Farmer, the chairwoman of the Martha Cook committee. She was a fellow law student of Joe's.

"Bina, we understand why you didn't sign up for the dinner, but if you would reconsider, I'd appreciate it. There's a real tall law student, a friend of mine, who lacks a partner. I'd hate to disappoint him."

"What's his name?"

"Joe Kiyonaga."

I assumed he was Italian, thinking the name was spelled something like "Chionnaga." I agreed out of kindness. I couldn't bear the thought of leaving someone in the lurch, having often been in that situation myself.

Across the street, Joe Kiyonaga had also failed to sign up for the dinner. He, too, was engaged, ring and all, to Fumi, a Japanese girl from Hawaii, attending Barnard. They'd met in Hawaii after the war. She was Japanese and stood at 5'1", with long, straight black hair and an affected blond streak in the front. She must have been pretty attractive—not only did she attend Barnard, she also represented the Hawaii Tourist Bureau at Rockefeller Center. She and Joe planned to marry the following summer.

The chairman of the Law Quad approached him, much as I had been. They described me as being tall, redheaded and without a date. Joe accepted, mainly out of curiosity. Tall girls were in short supply in Hawaii.

Thursday night arrived.

I'd always been a conservative dresser, operating on the principle that quality outweighed quantity. I usually had one really good outfit. I chose to wear it that night—a simply cut black wool-crepe dress and black snakeskin pumps.

Never one to wear makeup or curl my hair, beauty preparations played no part in my being late. I'm always late.

Martha Cook's drawing room was pretty impressive. It was cavernous and mahogany-paneled, with arched, leaded windows and lustrous hardwood floors waxed to a soft, buttery glow. A huge crystal chandelier dominated its peaked ceiling, and a Steinway baby grand held court in its far corner.

The other corner was every bit as commanding. There, alone, stood a tall, rugged, quietly exotic-looking man—Joe. He looked a little lost, as if he'd been stood up. But there was no question who I was. I was his date, as I made abundantly clear—crossing the vast expanse of the Persian-carpeted room with my hand extended.

I never was very good on blind dates. I always talked too much. This evening was no exception since I assumed that Joe spoke little English. Besides, I wasn't working too hard to make a great impression—being engaged, I considered myself more or less out of the running.

Joe, on the other hand, didn't seem to let his engagement stand in his way, and his English (once I stopped talking) was surprisingly good. But, good Lord, that tie! My eyes were riveted to his chest. His tie was green, emblazoned with palm trees. I couldn't imagine so attractive a man having such poor taste. I was less than diplomatic.

"I can tell that you are from Hawaii. It's written all over you."

"You mean the tie? Oh, I wore it as a joke," he laughed.

His humor was lost on me, but Joe got the message. Palm trees might be fine in Hawaii, but Michigan called for something more conservative.

Tie aside, there I was in the company of the most attractive man I was ever to know. Charming, witty, and eight years my senior, he seemed singularly un-Japanese. I was to learn otherwise.

I think the main reason that Joe and I got along so well is that we assumed nothing could possibly come of the two of us. It was simply outside the realm of possibility. I mean, a Japanese boy and a Baltimore girl? I'd never heard of it. We both just relaxed and enjoyed the evening, and each other's company.

At the end of the evening, Joe invited me to an upcoming law school dance. He said he had a lot of studying to do, so could I please pick up the tickets? I agreed. Right then the footing for our relationship was established.

≥≈

THE FOLLOWING NIGHT A GROUP OF GIRLS FROM THE DORM gathered in Nova Muir's room. Hers was a natural meeting place. She was well liked, vivacious and never lacked for suitors. I couldn't say as much for the rest of the group. While we were sitting around talking, Nova was called to the phone. A few minutes later, she came back and announced:

"God, what a relief. I was afraid I was going to have to kiss that Jap!"

"Who?" asked her roommate.

"Joe Kiyonaga. He canceled our date for next Wednesday. Apparently he's met someone."

Her roommate was amazed. "I thought you liked him."

"He was fine for an occasional date, but I'm well out of it."

Underneath her bravado Nova was seething. Men didn't drop her, she dropped them.

"Jap."

I'd heard the term used throughout the war. I'd even used it myself. I never referred to the "Japanese" and the "Germans." It was the "Japs" and the "Nazis" —the enemy! But this was different. This was Joe. I winced. I took offense for Joe and, I guess, for me. I didn't realize it at the time, but it was the beginning of an eternity of winces.

So Joe had canceled the date. That was Joe for you. Anyone else would have feigned sickness or pressing studies or any number of plausible excuses. Innately decent, Joe was simply being fair. He had met someone whose company he preferred. It's a good thing that Joe canceled. Once Nova had kissed "that Jap," she never would have let him go.

❧

SATURDAY NIGHT ARRIVED, AND WITH IT JOE HOLDING A single white camellia. He'd gotten the message all right. There he stood in a beautifully tailored gray flannel suit, white shirt, and subdued foulard tie. The evening was perfect. It was a dance buffet, and Joe turned out to be a smooth, understated dancer. As the evening grew late, he began holding me closer as we danced, and I seem to recall that he gave me a glancing kiss on my forehead. I backed away and said, "Please, not so mellow." I actually was beginning to feel guilty about what's-his-name in Bogotá.

So I tried to douse the mood. I'd noticed that Joe had a scar on his lower lip which ran halfway down his chin. I asked him how he got it.

"Well, Bina, you realize I was in the war, and saw a lot of hand-to-hand combat. One day, I was surprised by a German with fixed bayonet."

This was getting interesting.

"Really, what did you do?"

"I let him approach me, and when he did, I bit off the tip of his bayonet."

"You mean it?" (Talk about naive.)

"Uh-huh, and I've borne the scar to this day."

(In actual fact, Joe was in an automobile accident when he was very young. His head went through the windshield. There were no doctors available on Molokai, so his mother used her seamstress kit to patch him up as best she could.)

❧

THE NEXT DAY, A BEAUTIFUL AUTUMN SUNDAY, JOE AND I went for a walk around campus and wound up at the local drugstore. Joe was short of change, so we split a can of Campbell's chicken noodle soup and a coffee. Shortly after I returned to the dorm, Joe left a hand-delivered note at the front desk:

Bina,

> *Sunday is a wonderful day. It will always be a wonderful*
> *day because you have become a part of it.*
>> *Joe*

On our fourth date, Joe and I were heading across the campus after studying in the library. It had started to rain—a light sprinkle that you hardly notice until you're soaking wet. Joe pulled us off our path to the back of the drama school. (I think the rain had figured perfectly in his plans.) He held me in his arms and kissed me for the first time. When he pulled back, he looked at me and said,

"Bina, I love you."

I had just finished reading a novel where a girl responded to a similar declaration by whispering, "I know." In the book it seemed perfect. I looked at Joe, and murmured, "I know." I guess Joe hadn't read the same book.

"For God's sake, I take it back."

The next week, he picked me up late one afternoon to go for a short walk. Joe didn't have much money, so we took a lot of walks. Only a block into our stroll, he stopped, turned to me, took my hand, and proposed in three little words:

"It'll never work."

"Sure it will," I said.

We were, I suppose you could say, engaged.

❧

FOR ME IT WAS THE BEGINNING OF THE REST OF OUR LIVES. FOR Joe, it meant we could finally sleep together. Thus, his campaign to get me into bed proceeded apace.

"How do we know if we'll be compatible?"

"How do we know if we'll fit?" (Joe, the scientist.)

I kept my cool—and my virginity.

Joe lost his . . . cool.

One night, after another attempt to convince me that the best thing for both of us would be a tumble in the hay, he dropped

me off at my dorm. His frustration was palpable. He didn't kiss
me goodnight. Instead, he turned away from me, took a few steps,
paused, pulled a nickel out of his pocket and skittered it across
the sidewalk in my direction.

"If you change your mind," he said, "give me a call."

It was the first I'd ever seen the icy side of Joe. It was also
the last time I saw Joe for three weeks.

I did receive a letter from him the following day.

> *Bina,*
>   *Permit me to apologize for my behavior when I left you
> last nite. I am sorry about that as I am sorry for breaking the
> date for tonite. I don't plan to go to the dance or to Sherry's.*
>   *Good-bye.*
>   *Joe*

Without missing a beat, Joe started to date someone else. I,
on the other hand, proceeded to be miserable—and to get a tan!
I figured if he ever did see me again he'd sit up and take notice.
A tanned Bina's eyes were bluer, hair more red—even my freckles
managed to look good. I'd fix him. While I tanned, my "friends"
reported sightings of Joe with "Mary Hill," a tall, lovely blond
sorority girl—at dances, playing tennis, or sitting hand-in-hand on
the Law Quad steps. I was miserable, but busy.

I spent my early spring afternoons on our building's roof with
my friend, Connie V. We'd stash Cokes and chips and settle in
for the long haul.

One afternoon I received a call from the receptionist that "a
Joe Kiyonaga is here to see you."

I slipped on my pale blue pleated chambray skirt, a white
cotton shirt, socks and saddles and went down to meet him. He
took one look and he was back—on my terms. I never made
mention of his absence or the sorority girl.

⌗

BUT WHILE THINGS WERE ONCE AGAIN FINE BETWEEN JOE AND me, the rest of the world was another story.

Soon after I met Joe, I wrote my parents a letter about classes, campus life and Ann Arbor. I also, with studied nonchalance, mentioned that I happened to be dating a Japanese law student. Shortly thereafter, I received my one and only phone call from Bogotá. It was my mother calling to tell me that she and Daddy were flying in from Colombia the following weekend. She hardly needed to say why, but she did anyway—she "didn't like the idea of my dating a Jap."

I was both surprised and disappointed. First of all, we were hardly bluebloods who had to worry about our family name being sullied. Second, I'd never seen any sign of bigotry or racism in my house while growing up. And third, with a father as a diplomat living in Bogotá, I considered my parents to be cosmopolitan. But I didn't try to talk Mother out of a visit. I looked forward to seeing both my folks and to introducing them to my new beau. Joe wasn't too crazy about the idea; he thought it was too soon.

I reserved adjoining rooms at the Book Cadillac, then the fashionable hotel in downtown Detroit, and took a bus to the city. Joe made his own way to Detroit and stayed elsewhere. Dinner reservations were for the hotel's main dining room. After settling in, the Cadys went downstairs to the bar to meet Joe Kiyonaga.

It was a cold night, and Daddy and Joe, it turned out, were dressed much alike: both wore gray flannel suits, white shirts and rep ties. Mother was dressed fit to kill in a black dress, a mink fur and a hat with a formidable feather that I thought was pretty unbecoming. I completed the picture in my tweed skirt, gray cashmere sweater with a single strand of pearls, my snakeskin pumps and a black alligator bag. The stage was set.

We entered the rather formal dining room to find Joe seated at the nearby bar with his back to us, drinking scotch neat. I'm guessing that my parents expected to find a stereotypical-looking Japanese (I'd not given them a word of physical description). They must have been greatly surprised when Joe turned and uncurled himself from his bar seat to greet us. He stood an inch above my father's six-foot, three-inch frame.

Conversation was strained at the start, despite the fact that my father had had what we used to call a "dressing drink," a cocktail before leaving his room. Daddy ordered a round of old-fashioneds, and the mood began to thaw. My mother never had been socially assertive, so she remained relatively quiet as Joe and Daddy began to talk. Uncharacteristically, so did I. The men discussed Bogotá, the weather, life on campus, Hawaii—everything, in other words, except what was really on their minds.

Wine at dinner didn't hurt. In fact, with each sip, everyone seemed to look and feel better. I'd been wary, too. Like Joe, I'd thought Mother's rush to come north was precipitous. But both Joe and I began to relax as dinner progressed. While I certainly wanted my parents to like Joe, I wasn't agonizing over their verdict.

My mother had always told me when I was growing up to pray to Our Lady to help me recognize the man I was meant to marry. I followed her dictates to the letter. Each night for as long as I remembered, I'd pray to the Virgin Mary that I would know enough to recognize my future husband. I wasn't praying *for* a husband, just for sense enough to spot him when he showed up. I looked at Joe as he smiled quietly and warmed to my father. He'd shown up all right.

After dinner, Joe left the hotel. Mother was tired, so she and Daddy went upstairs. There was no mention of the evening until the next morning, when my father rang me and invited me to join them for coffee in their room.

I put on my robe, and knocked on the adjoining door. As I walked in, my father put his arms around me and said, "Bina, I like Joe. He's the only man who can handle you."

With that, Mother stood up from the breakfast table, embraced me, and added: "I think he's lovely, Bina."

The verdict had been swift and sure. Daddy didn't even mention George, whom he had really liked. Then again, there was only one Joe Kiyonaga.

❧

SOON AFTER THAT FIRST MEETING, I SPENT EASTER BREAK BACK in Washington, D.C. with my parents. Joe stayed on campus to study. We were both lonely. An excerpt from a letter that he wrote me:

> *The quiet is very loud here. Makes me feel like shaving the old pate and taking the oath. How do you think I would look in a smock?*
>
> *It's too bad you will miss the blossoming of the cherry trees. I had visions of you gamboling on the greens and feeling like Cio-Cio-san.*
>
> *. . . I miss you like the devil but I'm suffering in silence like the Christian martyr that I am.*
>
> *Give my regards to the mater and pater. I love their daughter.*
>
> > *Josef*

Convincing my parents that I knew what I was doing was nothing compared to the girls at Martha Cook, the cream of the Michigan crop. I remember two friends in particular, "Ginnie" and "Kathy," telling me that marrying Joe would be the biggest mistake of my life. "Where will you live? No nice neighborhood will have you. You might as well kiss the idea of belonging to any club good-bye. And what about your poor children?"

I'd come from Harlem Avenue; I'd never belonged to a club; I'd never had children. Marrying Joe seemed the right thing to do. For the first time in my life, I was comfortable.

❧

JOE AND THE LAW JUST WEREN'T MEANT TO BE. ON JULY 9, 1947, I received a pretty matter-of-fact telegram.

FLUNKED OUT. NOT GOING TO ANOTHER SCHOOL. WILL WRITE. JOE

Later, discouraged, he told me he'd "rather be a first-rate farmer than a third-rate lawyer." (I would have thought the options a little broader.) He'd decided to return to Hawaii. This was certainly a surprise, as we'd planned to get married after our second year at Michigan. Now, at least for Joe, there was not going to be a second year. He said he'd fly to Washington the next day.

I picked him up on a Tuesday evening at the airport with one question on my mind: Where did I fit into Joe's new plans?

I wasn't sure if I wanted to be part of his plan when I saw him get off the plane carrying an enormous cardboard box tied with frayed twine. (His regular luggage was also aboard, as it turned out). In the car, Joe told me he intended to ask my father for my hand, return to Hawaii, and we'd get married the following year. It seemed a little vague to me, but I didn't react at the moment. We went straight home.

My parents, Joe, and I had a drink, and then Joe asked to speak to my father alone in the living room. My mother and I went to the kitchen and promptly put our ears to the swinging door that connected both rooms.

Joe observed all proper conventions of the time in asking for my hand.

"Mr. Cady, I would very much like to marry Bina."

My father told Joe that, personally, he liked him, but he was worried about whether he could support me. (I'm sure the specter of past clothing bills from Pasternaks and Garfinckels flashed through my father's mind.)

"Well, sir," said Joe, "I've saved $6,000 and I hope to earn more. I also own some property in Hawaii."

The discussion continued for about an hour. From the way Daddy was carrying on, you'd have thought Joe was trying to marry into royalty.

Joe agreed to see me through my last year of college out of his meager savings. To this day I am surprised at that stipulation, as Daddy was a generous man and fairly well fixed. Besides, both Mother and Daddy seemed more than happy at the prospect of having me off their hands. When Daddy suggested that I finish college before marriage, Mother turned to me and objected strenuously. She told me that was a bad idea. She sensed that Joe might be a little "skittish" and thought it better to "strike while the iron was hot" (a less than apt metaphor in my case). She also suggested that, once out of sight, Joe would be "snapped up" by someone else.

Eventually, Daddy called Mother and me back into the room and announced that Joe had asked to marry me and that he'd given his consent. My mother acted surprised and delighted. But I was still mulling over the one-year part of the deal.

Joe and I headed out to dinner by ourselves. Once in the car, I raised Mother's concern.

"Joe," I said, "Mother said she's afraid you're skittish."

"What do you mean?" he asked. And I repeated Mother's concerns word for word.

Joe threw the problem back on me: What did I suggest we do about it? Without hesitating, I replied, "Let's get married Friday."

Joe was even more stunned than I was at my proposal. He'd arrived in Washington thinking he was a year away from getting married. Now he was confronted with the prospect of facing a priest in a few days.

Not to be outdone, Joe said, "Sure." Barely audibly.

At dinner, we agreed that he would return to Hawaii to find a job, and I'd follow a few weeks later and begin my senior year at the University of Hawaii.

The next morning, Wednesday, Joe and I went shopping for my wedding band. We visited the most reasonably priced jewelry store in town, Charles Schwartz Jewelers in Washington. On the way, Joe promised that when I got to Hawaii, he'd have a friend who was a jeweler make me a wonderful wedding ring, studded with rubies and diamonds. At Charles Schwartz, we were just shopping for a temporary band.

But Joe neglected to explain this to the salesman as tray after tray of wedding bands were uncovered for our examination. They were all too expensive. Finally, we got down to basics, and the salesman proffered a simple gold band. It was $15.00.

Joe countered: "Could we settle for $10.00?"

The salesman shot me a pitying glance. With a sigh he handed it to Joe, and said: "Take it."

I still wear this wedding band today. I never would have considered a substitute. (I think.)

Joe didn't wear a wedding band. When I asked if he wanted one, he said no. "I'll know I'm married."

I'm not sure when or how Joe broke the news of our engagement to Fumi, but I don't think she took it very well. When I asked Joe if I could see the engagement ring, he told me that that wouldn't be possible. She'd never returned it.

᪥

ON THURSDAY, JOE AND I WENT SHOPPING FOR MY WEDDING dress at Hutzler's in Baltimore. I'd worked there during high school, and it was considered the nicest department store in town. Afterwards, we lunched at Millers, a restaurant known for its oysters and mint juleps (not necessarily together). I could tell Joe had something on his mind.

"Bina," he said, "I have an idea. Now that I've decided to chuck law and go back to teaching in Hawaii, we won't have much money, even a place to live. Don't you think the Church would allow birth control for a few months?"

Reluctantly, I ventured, "I suppose we could ask."

There was supposed to be a really bright young priest who had served as a chaplain in the war and had just arrived at Saint Ignatius as a resident. After lunch, we drove to Saint Ignatius. The Jesuit rectory, in the heart of Baltimore, was a forbidding, puce-colored structure that looked as though it probably housed the overflow from its adjoining cemetery. I sure could have used my roller skates and whistle right then.

"Joe, why don't you come in with me?"

"No, no," said Joe airily, as he pulled out the newspaper. "You go ahead, I'll take your word for it."

The thick rectory door—straight out of *The Addams Family*—

creaked open to reveal a wizened, crouching priest. Oh, no, surely he wasn't the one I wanted—he could hardly be abreast of modern Catholic thought.

"Good afternoon, Father. I wonder if you have a resident theologian?"

All Jesuits consider themselves theologians. But if he was offended, he didn't show it.

"Come in, my child," he said understandingly. "Wait in the parlor and I'll fetch Father Mahoney. He's just today back from a lecture tour."

I knew I'd hit the jackpot.

A few minutes later the parlor door opened with a flourish upon a handsome young Jesuit.

"Good afternoon, my child," he said even though he wasn't much older than I. "How can I help you?"

"Good afternoon, Father. I promise I won't take up much of your time."

"No, indeed, your visit affords a welcome break in my schedule."

He'd been trained as a speaker, all right, and probably had been a comfort to the troops as well.

"Father, I'm about to be married."

"Go on, my child. Tell me, who is the fortunate young man?"

"His name Is Joe Kiyonaga. He's from Hawaii."

"Kiyonaga? What kind of a name is that?"

"It's Japanese, Father, and I'm looking to you for guidance concerning birth control."

Father Mahoney closed his eyes, held up his hand, and with a pained expression, said, "No, no, say no more. I understand perfectly. Marrying a Japanese, having children would be out of the question."

I got up and left.

Back in the car, Joe never looked so good.

He put down his newspaper.

"Well, what's the word?"

I smiled at him.

"No birth control."

﹏

WE WERE MARRIED THE NEXT DAY, JULY 25, 1947, AT SAINT
Matthew's Cathedral in Washington, D.C., with only Mother and
Daddy in attendance. Mother arranged the entire wedding in just
three days. She even managed to avoid the Church's usual prac-
tice of announcing the wedding banns. She was dealing with a
Latin priest, and he greased the ecclesiastical skids.

The weather couldn't have been more glorious. I wore a short-
sleeved pink linen suit with fresh flowers in my hair. Joe remarked
later that he wished I'd worn stockings for the occasion, despite
my tan. I was just too relaxed.

At five in the afternoon, Joe and I entered Saint Matthew's
by way of the sacristy. Father Augusto Trujillo waited for us at
the main altar. Joe and I stood just inside the altar rail and were
married. When we turned to walk down the aisle, I was surprised
to see that the church was far from empty. People had dropped
in on their way from work to make a visit and attend 5:30 P.M.
Mass. They were our captive wedding guests and shared in our
joy. As we made our way down the aisle, we nodded and smiled
at them in turn. One man stood up and shook Joe's hand.

Daddy took us to Harvey's for dinner, known as the "Restaurant of Presidents." The Catholic Church was still in the era of
meatless Fridays, so we all ordered pompano "en papier." Daddy
lent us his old green Hudson and we were off to honeymoon for
a week in Williamsburg, Virginia.

﹏

I DON'T HONESTLY THINK THAT JOE AND I HAD ONE REALLY
serious conversation before we married. He still just wanted to
get me into bed. The fact I was a virgin served us well. Joe was
eight years my senior, had played the field, and was ready to settle
down. I was inexperienced but eager to learn.

As we were driving out of Washington, I informed Joe that I
thought that I was about to have my period. He drove as one pos-

sessed (Mario Andretti pales in comparison), begrudging me even a comfort stop—and there were many. As for losing my virginity, I was not nervous but I was curious. Early on I had accepted my mother's sketchy advice on womanhood. When, at age thirteen, I'd asked what this business was all about, she'd told me some rudimentary facts but suggested that I discuss it with no one other than my husband, and only once married. Joe had no idea what he was in for.

So unschooled was I in the ways of sex that I wasn't even sure how a man was built. I had no brothers. I had seen Daddy once in the act of putting on his underpants. I thought he looked kind of strange, but I blocked the thought completely from my mind since Mother had told me to "save myself for my husband and to give matters of sex as little thought as possible." I even averted my eyes in museums when I'd pass a statue of David or an occasional archangel.

As we raced through the darkness on the four-hour drive to Williamsburg, I looked at Joe with something approaching awe. He was dressed casually in a cream pongee shirt and matching whipcord slacks. I swear I never saw a man so handsome (nor have I since). I marveled that he had chosen me. I wore a comfortable two-year old black linen shift and flats. We made a good-looking couple. It's a great feeling.

We arrived in Williamsburg at midnight. Curiosity seemed to whet my appetite—for food.

"Joe, it's Saturday now. We can eat meat. Let's stop for a hamburger."

Looks may not kill but I felt a twinge.

I got my hamburger (cheeseburger, actually, with relish) and a malt. Joe refused all food (not even a taste). He wouldn't even have a soft drink. He sat glaring at me, drumming his fingers on the table. Probably finances again, I thought.

The Williamsburg Inn was the perfect choice for our wedding night. It's a beautifully proportioned colonial structure straight out of *Gone With the Wind* that combines a perfect mix of gentility and quiet comfort. To my mind, there's no match for Southern charm, and the Williamsburg Inn is its embodiment.

Mr. and Mrs. Joseph Kiyonaga approached the desk with less

than their usual bravado—me, because I wasn't used to my new name or status; Joe, because he wasn't certain how a Japanese would be received. We were received graciously and shown to our wedding chamber. It had twin beds, but otherwise it was lovely.

Joe proceeded to bathe while I unpacked. Truly, I was not a bit nervous, just very happy. My groom emerged from his bath, resplendent in burgundy and white challis pajamas. He'd bought them at Garfinckels as a special concession for our wedding night. He later told me he ordinarily dispensed with such convention.

Now it was my turn to dress for bed. As usual I took my time with my bath. Joe knocked on the door. Was I all right? I took the hint and donned my wedding nightie, a sheer ivory chiffon with a scanty lace bodice and cap sleeves. Then I topped it all off with a man-tailored foulard robe. (I really wasn't with it.) I was radiant, freshly scrubbed. I was ready to know my husband.

Joe took one look at me and said,

"Take off your robe."

I did.

With that, I had my first sexual encounter. It lasted seven minutes, tops. Satisfied, Joe fell on his bed in a heap and slept. I guess after all those months of keeping him at bay, that cheeseburger (not to mention the leisurely bath) had just been too much.

Watching him sleep the sleep of the unjust, I thought, "What a pig! And he didn't even have the decency to kiss me goodnight."

I now knew what a man looked like all right—and what he felt like, as a matter of fact—but I failed to understand why the commotion. In the bathroom, as I washed up, I realized that I was bleeding.

Saturday found a repentant Joe. He must have understood that the night before hadn't been exactly the way I'd envisioned our first night together. He made up for it in the morning. When I woke up that afternoon from our nap, Joe was nowhere to be found. But his touch was evident. He'd left a single red rose on his pillow.

I had told Joe before that I loved him, but I was just beginning to realize what that meant. I adored him, and I wanted to tell him so. So I did.

His reaction was strong. I was never to say that again, he

cautioned me, because he could never say it back. And why not? Once said, he would no longer be his own man. Maybe it was a Japanese thing.

"All right, Joe," I said, "don't bother to say it to me, but is it all right if I say it to you once in a while?"

"Fine, Bina, but don't expect anything in return."

Joe, the magnanimous.

⁂

AFTER THREE DAYS OF WILLIAMSBURG, JOE HAD HAD ENOUGH of Colonial atmosphere and short beds. He longed for the sea. Me, I'd seen nothing but the inside of our room—and Joe, of course—since my arrival. I suggested a night on the town. The big attraction in Williamsburg, other than William & Mary, is the Governor's Mansion. We went for a visit.

Dressed for the occasion, we took off, hand in hand, down the main street of Williamsburg that ended at the mansion. It was an impressive sight. Candles glowed in every window and kerosene torches dotted the high wall surrounding it. At the gate we were confronted by a guard. The charge was 50¢ per person.

Joe was indignant. Why pay to see a public edifice? The guard patiently explained that the edifice in question had been restored to its original state at considerable cost, thus the tariff.

Joe's solution was simple. "Once you've seen one governor's mansion you've seen them all." (I'd never seen a governor's mansion.) "We can see all we want from right here."

With that Joe proceeded to stand on his toes and peer over the wall. His height was a distinct advantage. I tried peering. All I could see was the roof. But I'd just about had it. Stalking back down the main street of Williamsburg alone, I couldn't help but think, "Loving aside, what kind of chintzy creep have I married?"

Joe was fast on my heels, feigning confusion. He knew why I was mad but he came up with another solution. Why not check out of the Williamsburg Inn and move on to Virginia Beach? At least there we could enjoy a swim rather than all the "phony baloney" sightseeing.

I chose to ignore the sightseeing remark. We were off to the beach. My glowering presence next to him in the car occasioned Joe's request for a divorce. Astounded, I realized how greatly our views of marriage differed. Mine was a total commitment. His, apparently, was one of a series of three-night stands.

I didn't hesitate as I turned to answer him. "Joe, we are married. M–A–R–R . . ."

"I can spell."

"Fine, if you're such a great speller try looking the word up in the dictionary. You'll find that we are man and wife not for three days, three months, three years. We're husband and wife forever. Just try leaving me and I'll dog your footsteps; I'll hound you; I'll follow you to the ends of the earth, and if necessary I'll make your life hell; but we'll stay married. I'm your wife, Joe, and please don't ask for a divorce again."

He never did.

He'd met his match.

And I'd met mine.

❧

CHAPTER IV

ALOHA

Six weeks.

That's a long time to go without seeing your husband. Especially when you've only been married seven weeks.

It was September 13, 1947, as the Pan Am Cabin Clipper circled Oahu, Hawaii, on its approach to the Honolulu Airport. From the window I could see the lush green valleys enshrouded in mist, the talcum-powder beaches, the aqua blue of the Pacific glinting at me in the late afternoon sun. Hawaii was saying hello.

I brushed my hair and pinched my cheeks (I still wasn't into makeup), and straightened my yellow linen dress. I remember it well. It was one of my favorites—simply cut with a "Peggy" collar and bamboo buttons down the front. I was ready to see my groom.

❧

MOTHER'S PARTING WORDS WERE STILL WITH ME. BEFORE I'd left, she'd taken me aside for some motherly advice:

"First, never learn to sew."

"Second, always have help in the house."
"Third, keep some money set aside for yourself. You never know."
(Mother'd always had a practical bent.)

❧

AS THE PLANE TAXIED DOWN THE RUNWAY, I ENVISIONED OUR first few nights. What had Joe planned? Would he whisk me away from the airport to some secluded beach? Dinner at some torchlit oceanfront restaurant? Or maybe we would go home—to our new home—and just be together.

As I stepped off the plane, I caught my first glimpse of him. You couldn't miss him. I'd like to say our eyes met. But Joe was too busy, surrounded on the tarmac by a large group of Japanese. He approached, we embraced quickly and he gave me a brotherly kiss. Introductions were in order.

I met about twenty people. It was hard for me to tell them apart—Orientals and nuns always looked alike to me. I just smiled and did my best to look comfortable. Last, I met Joe's mother. She was formally dressed in a black silk suit and white blouse.

Her reaction was silence. Dead silence. I was too tall, too redheaded, too white. She took one look at me and the battle lines were drawn. Better we had remained "pen pals."

Joe hadn't prepared me for any of this. While we were dating, Joe had shown me a small black-and-white photo of his mother. She was standing by a fence holding a small dog. She looked nice enough.

As I waited for my luggage, surrounded by Joe's friends, I paid little attention to the curious stares of passersby (and there were many). I wasn't quite sure what was happening. But I knew one thing: I'd been in Hawaii about three minutes and already knew it would never be my home.

Joe suggested dinner and invited his mother along. It struck me as a little odd, but it was fine. He selected the Wagon Wheel: touristy, nice, but hardly Hawaiian. (So much for grilled prawns under a full moon.) I think it was the least Japanese place Joe could come up with. He must've thought it'd be like a touch of home for me.

The place was downright creepy. It kind of put me in mind of our basement in Baltimore (now there's a touch of home for you). We were ushered to a very quiet, very dark corner table next to a giant wagon wheel—they were everywhere. I couldn't help but feel like a pioneer. I suppose, in a sense, I was.

Joe acted as if we were on our first date, with his mother as chaperone. I think we actually talked about the weather, a singularly uninteresting topic in Hawaii where the weather is virtually constant. I tried to include Joe's mother in the conversation, and she seemed pleased when I called her "Mother." I recall that she spoke barely a word throughout dinner. She stared at me the entire time. As I ate, spoke, wiped my mouth, her eyes were glued. By dessert, I was quiet too. I was too preoccupied, trying desperately to ignore the feeling that I was sinking quietly into a black hole with no end in sight.

My body didn't ignore it. I woke up the next morning completely covered with hives. I had never had a single rash in my life. And I was scheduled to go with Joe's mother to a luncheon.

Hives or no hives, I was determined to be a good sport. Joe's mother said nothing about my rash. In fact, she said nothing. At the luncheon, she seemed more uncomfortable than I did. I didn't understand it—this was an attractive, interesting group of women who represented the upper echelon of Japanese society on the Islands (which, at the time, wasn't saying much, though the status of the Japanese would later change). I enjoyed getting to know them, but Joe's mother didn't really seem to fit in. I felt a little sorry for her.

⤝

THIS GROUP WAS KIND TO ME DURING OUR ONE-YEAR STAY *in Hawaii. Although I saw these women infrequently, they went out of their way to always make me feel welcome. In fact, the day before Joe and I were to leave Hawaii, they gave another luncheon for me. Their farewell gift for me was a lovely two-piece navy and green print "redingote" (a dress and full-length matching coat) to wear on the plane. I thanked them for the lunch, the ensemble, and, most especially, for their friendship. But I couldn't leave well enough alone. I ended my speech, not really thinking, by saying:*

"As much as I've enjoyed your company, I'm not sure I'd like it as a steady diet."

Good thing I was already wearing the outfit.

❧

I HAD THOUGHT HAWAII WOULD BE THE LAND OF MILK AND honey. Instead, it was more like sukiyaki and rice. Japanese were everywhere I looked. We lived in a Japanese neighborhood, ate Japanese food. Everyone spoke Japanese—and I was odd man out. I felt like a caged animal, and there was no escape. No whiff of home; no Uncle Timmy; no oysters on the half shell; no belly laughs—no fun!

God, what had I gotten myself into? Even Joe seemed different. It was one thing to see him on the Michigan campus; there he looked exotic, distinguished, even. Once we got to Hawaii, he reverted to type and became—Japanese! Deferring to his mother like mad, treating me like some insecure interloper (which, as a matter of fact, I was). I knew the war was over, but I was beginning to hate the Japanese all over again.

The girls at Martha Cook had been pretty much on the money when they'd warned me of the obstacles I'd face if I married Joe. I wasn't only marrying one man: I was marrying into a whole new world, a world that became less exotic the closer I got; a world that seemed just plain strange.

As a new bride in Hawaii, I had a recurring nightmare. It was always the same. I'd be back at the farm, heading up the lane toward the Manley house. Just as I'd open their gate, the house would sink out of sight. A huge bottomless void would open in its place, at my feet. At that moment, I would lose my footing and, as I'd try to catch myself, I'd be pitched into awful nothingness. The dream would end with my cry for help. That's when Joe would hold me in his arms and comfort me.

✌

DEALING WITH A MOTHER-IN-LAW IS TOUGH; DEALING WITH A Japanese one is an absolute disaster. In Japanese culture, a mother-in-law (the man's mother, that is) has the unparalleled, and virtually unchallenged, position of in-house food critic, household management advisor, and child-rearing guru.

Oahu was a short flight from Molokai, and Joe's mother was a frequent flyer. What's more, she didn't like to travel alone. She had a tendency to drop in unannounced and bring several friends, which was about as appealing as a spot check by a team of IRS agents. I'd come home from school (I had enrolled at the University of Hawaii) on Friday and there they'd be, all set for the weekend.

Joe's mother had a way of taking over. Everything I did (especially in the kitchen) was wrong. She was in her element. Laughing, she'd hustle her way into the kitchen and point out my mistakes to Joe and any friends whom she'd brought along. When you're far from home and have never been in a kitchen, this sort of thing looms large in your mind. I'd just leave and let her cook. (And she was an excellent cook.)

✌

THIRTY YEARS LATER, JOE'S MOTHER CAME TO VISIT US IN *Chevy Chase for Thanksgiving. I'm a great one for Thanksgiving: homemade oyster stew, crab ravigote, a fresh turkey with apple/pecan stuffing, the whole spread. It was a chance for me to make a nice meal. I'd come a long way in that department.*

Joe's mother had some special dishes she wanted to make. Fine. Five—no, maybe six—hours later, she was putting the finishing touches on her coconut pudding made from scratch (more of a paste, actually), sushi (who eats sushi for Thanksgiving?) and other

Hawaiian-inspired delicacies. The oven was finally free. It was time
for me to roast the turkey. The guests were due in two hours.
The turkey that night could have passed for sushi.

❧

I UNDERSTOOD FROM JOE'S MOTHER THAT I WAS TO WALK FIVE
paces behind my husband, speak only when spoken to and, should
I become pregnant, I was to give our first child to her—a so-
called custom in ancient Japan. Here I was in the middle of nowhere. Friendless. No one to talk to. Was I to accept the dictates
of Joe's mother? Either I was in the wrong pew, or the woman
was nuts!

The worst part of my newly married status was Joe. He
seemed to side with his mother. He'd walk with her, and I'd be
left to bring up the rear. He'd talk to her in Japanese, so that left
me out. I did put my foot down when it came to whose child
would be whose.

Maybe it wasn't anyone's fault. This was new for all of us.
And his mother, in her own way, did try. She gave us our first
wedding gift: a sparkling new Singer sewing machine. (It seems
she hadn't conferred with my mother.)

She wasn't cruel, she was threatened. I don't think she disliked
me, just everything that I represented. She took one look, and
knew I would take her son away from her—and from Hawaii.

I like to think of myself as being a Christian. I've always believed that the gravest sins are the ones committed against charity—and here I was being unkind to Joe's mother. She was, after
all, his mother. To be honest, I was ashamed of her. It was bad
enough to feel that, but I communicated my feelings to Joe
which, in turn, converted his unease into pure discomfort. So we
were all miserable.

I mentioned before that I could've learned a valuable lesson
from my own mother, but I didn't. But I did try.

❧

IT WAS MY BRAINSTORM TO VISIT JOE'S MOTHER IN MOLOKAI.
I thought it would be a sign of respect to her. Joe resisted. The
man really didn't want to go. I insisted.

The flight may only take thirty minutes, but it takes you thirty
years into the past. Our prop plane touched down on a desolate,
silent, dirt runway more akin to someone's backyard. More dry
and barren than Oahu, Molokai had a weathered beauty. I felt
like I'd landed on the moon.

We were off to town—Kaunakakai—over a bumpy, dusty, un-
paved road. Kaunakakai's main street looked like an old Western
movie set. One block long, it housed a grocery, drugstore, the
Midnight Sun (or, as the locals called it, the "poi palace"), the
Bon-Ton Dress Shoppe (Mother's shop), Iwamura's Variety Store,
a post office, a police station, a one-room wooden cellblock, and
Leilani's Bar—the most popular spot in town. All fronted a dirt
road with a makeshift sidewalk.

Two miles out of town we rounded a bend, and I immediately
understood why Joe hadn't wanted me to make the trip. Set back
from the road amidst trees and underbrush was a lone wooden
house. The place was straight out of *Tobacco Road*. Truly. It was
shingled, held up by wooden supports. Chickens pecked at my
ankles as I retrieved my bag. Never squeamish, I was just in shock.

Inside, it was small and simple. I was interested to see that
Mother's bedroom housed a small Shinto shrine (despite the fact
that she was a practicing Catholic) with a picture of Junzo surrounded by fruit, flowers and candles. She was really trying to
hedge her bets.

The evening was strained, but Mother did what she could to
make us comfortable. She'd made all of Joe's favorite dishes, even
opihi (raw sea barnacles) and poi. I was just grateful that she
didn't serve chicken. Later Joe and I went to bed—a single cot
on a side porch.

Our plans to spend the weekend changed after breakfast the
next morning. I asked Mother for a piece of bread, for toast. (I
was a little queasy—must have been one sea barnacle too many.)
She handed me a slice of white bread, peppered with ants.

Joe suggested that we leave on the next flight. Throughout our thirty-year marriage, that was the only time I set foot on Molokai.

❧

MOLOKAI MAY HAVE BEEN A NIGHTMARE, BUT OAHU—WHEN IT was just Joe and me—was starting to grow on me. I was getting used to things, and how could I help but like the island itself? It combined perfect weather and beautiful surroundings. (Millions of people don't flock to Hawaii for nothing.)

Early during my stay in Hawaii, Joe and I went shopping for groceries. I was heading down an aisle, when Joe abruptly grabbed my elbow and tried to steer me to an adjoining aisle.

"But, Joe, we need soap."

"That's where we're headed," he objected.

I forged ahead, undeterred, toward the blue-and-white boxed Ivory Flakes. But before I'd taken more than a step, I heard "Yoshio, Yoshio" (Joe's Japanese name) being called out in happy recognition. I followed the call to its origin and spotted a shortish, fulsome Hawaiian young woman—made more fulsome by an obvious pregnancy.

Joe introduced us—this was Lei of the "lithe, dusky (he was right on that score) Hawaiian maiden" fame.

Just Joe's luck!

❧

WE LIVED IN A BUNGALOW JOE HAD INHERITED FROM HIS father on the fringe of Manoa Valley on McCully Street, a ten-minute drive from downtown Honolulu. It was small and had wonderful cross exposure—it reminded me of a treehouse. We had frangipani, white ginger, and oleander in the front yard. That scent! We were also surrounded by papaya, mango and avocado trees—breakfast, lunch, and dinner at the ready. It was a long way from Baltimore.

I was in my senior year, studying speech, at the University of

Hawaii. Joe was assistant principal at McKinley High School, teaching returned veterans who wanted a high school diploma. We'd meet each afternoon and head off to Waikiki. Our favorite spot was in front of the Halekulani, now a pricey luxury hotel, then a rustic collection of beautifully appointed bungalows. That's where the surfers (many of them Joe's students) hung out. They'd take their huge, wooden surfboards, paddle out a hundred yards or so, and catch a wave. Occasionally, one of the students would take me with him. I'd lie down on the board as he stood behind me. I loved it.

Evening walks became our ritual, although I can't honestly remember what we talked about. It was more just the feeling. We'd amble together at dusk along the Alo Wai Canal toward Waikiki, buttressed by Diamond Head, soothed by the steady surf. No skyscrapers. No tired throngs of visitors. No restaurants advertising breakfast specials. Just a few clusters of small hotels. It was old Hawaii. Life was simple, and, when it was just the two of us, we were happy.

Joe's mother aside, I'm surprised at the ease with which Joe and I began settling into married life. I guess neither of us had high expectations. After all, my parents' life together was steady but unexciting, and Joe's parents' marriage was a disaster. So anything was bound to be an improvement.

Joe and I spent almost all of our time together. I don't know if our isolation was intentional or happenstance. We just weren't in any rush to make friends; I was enjoying getting to know my husband. Besides, whatever friendships we did make, I'd sometimes put in jeopardy.

❧

AT ONE POINT, I MADE FRIENDS WITH A JAPANESE NEIGHBOR who had a darling dog, a Scottie. She told me his name was "Brackie." Over time I'd pass my friend's gate, and call out to her dog, "Hi, Brackie!"

Joe was with me once and, to my surprise, got highly irritated. It seems he thought I was "making fun of the woman."

The dog's name was "Blackie."

❧

I WAS TWENTY-ONE YEARS OLD, FRESHLY MARRIED TO A JApanese, and living smack in the middle of the Pacific. The whole world was taking on new dimensions. Even something as ritualized as a wedding could take on a sublime and novel twist.

One of Joe's students, betrothed to a Hawaiian girl, invited us to their wedding. That Saturday evening, Joe and I headed down McCully Street, hand in hand, thereby violating my mother-in-law's number-one edict.

The wedding was at the bride's home, a simple frame house near the water, festooned with scarlet, yellow and pink bougainvillea. This was to be a real Hawaiian wedding, with all the men in aloha shirts and the women in muumuus.

We were greeted with what I was to learn was the Hawaiian wedding song. It was slow, stirring, almost plaintive. The groom began singing from the far reaches of the house—he had a beautiful voice. The bride answered him, singing from the opposite side of the garden. Then they came toward each other. It was one of those moments.

Then the party started, my first luau. Suckling pig roasted all day underground, lomi-lomi salmon, coconut cakes, mai tais, and mounds of fresh mangos and pineapples—all served amidst the gentle strains of the ukulele.

About a month after the wedding, I asked Joe to accompany me to a doctor's appointment. I didn't tell him the reason. I rejoined him in the waiting room, and gave him the news. It was the first time that I saw him cry. He was going to be a father.

❧

I WASN'T SURE HOW MY OWN MOTHER WOULD RESPOND TO my being pregnant. After our wedding, she'd taken me aside for a heart-to-heart. She cut to the chase.

"Bina, you're not pregnant, are you?"

"No, Mother."

She was visibly relieved.

"Oh, thank God. It's hard enough telling people that you married a Japanese, let alone having to tell them you're pregnant, too."

⁓

HAWAII WAS BECOMING OUR SIDE OF PARADISE, BUT IT STILL wasn't enough. We wanted more. We wanted out. For me, it was Joe's mother, and that "island fever" business was setting in. But for Joe, it was that and more. He was too good-looking and had too much promise to escape the jealous questions about his origins. Hawaii wasn't going to let him get away with it.

Meanwhile, I was still clueless about the controversy over Joe's background, though people did try to clue me in.

On my first visit to the dentist, he asked me, between fillings, to repeat my last name. I enunciated "Ki-yo-na-ga," thinking he was having trouble with the pronunciation. He seemed oddly interested, and asked if I was married to Joe Kiyonaga. I replied that I was. With that, and a look of quasi-amusement, he declared:

"Well, you know he's adopted."

On another occasion, we were entertaining some of my classmates at our home. This was the first time they'd met my husband. A male classmate from the drama department studied Joe up and down. He took me aside before he left and commented on what a striking-looking man I'd married. I was in complete agreement. He then asked what I understood Joe's background to be. (From the way he'd been staring, it seemed to me that his interest in Joe went beyond ethnic curiosity.) When I told him that Joe was all Japanese, he shook his head. I shuddered at his choice of words:

"Bina, there's a nigger in the woodpile somewhere."

Joe couldn't win. Either people said he wasn't all Japanese, or discriminated against him on the assumption that he was.

≫

A CLASSMATE IN SPEECH AT THE UNIVERSITY INVITED ME TO A luncheon at the Outrigger Canoe Club on Waikiki, founded years back by the famed Duke Kahanamoku. It was an unpretentious club for Hawaiian royalty, visiting luminaries, and *haoles* (the Hawaiian word for Caucasians).

Joe dropped me off at noon; he was to pick me up at two. I had a fine time at lunch with seven of my classmates. Two-thirty arrived, but no Joe. I began wondering where he was and decided to look around. I found him seated right outside the front entrance to the club on a makeshift rattan stool. When he spotted me, he stood up, brushed off his khakis, and asked:

"All set?"

"Joe, why didn't you come in? Why wait out here?"

"Bina, no Japanese are allowed."

≫

MAYBE IT'S NO SURPRISE THAT JOE DECIDED HE WANTED TO RE-turn to the States to study. He began considering his options. The University of Hawaii offered to send us to Yale, where Joe would work on his Ph.D. The catch was that after he received the degree, he'd have to commit to five years of teaching at the University of Hawaii. Even so, getting a Ph.D. and New Haven sounded pretty good to me. But Joe turned it down. Five years was a long time.

Instead, Joe chose to apply, at my father's suggestion, to the Johns Hopkins School of Advanced International Studies (SAIS) in Washington, D.C. Money was a concern, so Joe took a second job selling life insurance at night for John Hancock.

While Joe was moonlighting, I was busy studying and awaiting our first child. Pregnancy for me was a natural, simple affair. I didn't read any baby books. I didn't change my diet (or need to), and my maternity wardrobe consisted of oversized aloha shirts and shorts (except for a tentlike brown outfit that Joe's mother made for me). I walked a mile to and from school each day and continued riding the waves with the surfers.

I was well overdue when my young obstetrician, right out of med school (big mistake!), told me that Mary was breech, an unusually difficult delivery. That explained why he'd taken so many X-rays, but it failed to explain why he didn't stick around for Mary's delivery after I arrived at the Kapiolani Maternity Center. A simple wooden cottage tucked into the green hills of Oahu, it was a pretty place to have a baby. But scenery counts for little when you're in labor.

I was on the delivery table when the two nurses in attendance panicked. The last thing I remember, before they put an ether mask over my face and knocked me out, was the horrifying sight of the two of them sitting on my legs to delay Mary's arrival until the doctor returned. It's a miracle that Mary survived. They not only put me to sleep, but Mary as well. Thank God that she weighed in at eight pounds, six ounces, because she refused to eat for the first three days of her life. How could she? She was still fast asleep. I didn't fare much better. I was in pretty bad shape and was cautioned against having more children for fear I couldn't carry them to term.

Despite her rough entrance into this world, Mary was beautiful, with reddish-blond hair in front and black in the back. (I figured that's what came of a mixed marriage!)

Soon, we got the news that Joe had been accepted at SAIS. So on September 13, 1948, exactly one year to the day of my arrival in Hawaii, we left for Washington.

I'd arrived alone; we'd become a couple. Now we were a family.

❧

CHAPTER V

THE AGENCY

THE IRON CURTAIN HAD DESCENDED. THE BERLIN AIRLIFT WAS in place. The Rosenbergs had betrayed us. China had fallen to the Communists. The Cold War was now in a deep freeze. It was 1949 in Washington, D.C.

So much for the frangipani and bougainvillea.

When Joe presented himself at SAIS, he was told he didn't have enough credits to enroll. Joe was incensed—while in Hawaii, he'd received word that he'd been accepted. He told the registrar he'd flown all the way from Hawaii with a wife and baby daughter, and he wasn't about to take "no" for an answer. Enroll me, he said, and if I don't do well, you can throw me out. They relented.

So Joe was in at SAIS, but *we* were practically out on the street.

❧

INITIALLY, WE STAYED IN MY FOLKS' SIXTEENTH STREET apartment. The place was ours until Daddy finished his tour of duty, this time in Brazil. We enjoyed the apartment and our neighbors as well. One of them was a young man who was just starting to make a name for himself as a criminal defense lawyer in the D.C. courts. Joe and I would walk to Mass at Saint Matthews with Edward Bennett Williams and his wife, Dorothy, and sometimes get together for a drink afterward. I enjoyed their company, but Joe got a little tired of Ed—"all that guy ever wants to do is argue." This from someone who'd just flunked out of first-year law school. But I guess Joe had a point. As you may know, Ed went on arguing (rather successfully) for the rest of his life.

Another of our neighbors introduced us to Tim Connally, a real estate broker. He came highly recommended. We met. He knew of a great, spacious apartment on Seventeenth Street, just one block from my folks' apartment. Off we went, with Mary in tow. I was surprised to see how different two streets could be, just a block apart. At that time, Sixteenth Street was pretty prestigious and Seventeenth was real seamy. We arrived at the rental apartment. It was on the second floor; roomy but grim.

Tim queried Joe on his last name.

"Kiyonaga? That sounds Hawaiian."

"Actually, it's Japanese."

With that Tim clapped his hands and beamed:

"We're in luck. Downstairs you have the perfect neighbors."

Downstairs housed a Chinese laundry.

"They're a lovely couple. You'll have a lot in common. Besides, they're great cooks."

That was the end of Tim—and Seventeenth Street.

After that, every weekend, Joe and I would pack Mary into Daddy's car and canvass the area. We went all over D.C., but the response was always the same: "No Japanese" or "No vacancies."

We shouldn't have been surprised. World War II had just ended, and Pearl Harbor was no dim memory. Feelings still ran high against the "Japs," as a baby-buggy saleswoman at Woodward & Lothrop put it, questioning my last name.

Joe was far from alone. I recall hearing that Dan Inouye, who had lost an arm in combat, couldn't get a haircut in California after the war—and he was in uniform at the time.

Joe didn't seem bothered. Then again, the Japanese rarely

show offense; they just prove their worth. Though we never discussed it, I knew he was feeling sorry for me while I was busy feeling sorry for him.

He told me not to worry, but I had my doubts. I even got to the point where I'd stay in the car with Mary, fearing that the sight of a Caucasian wife toting a one-year-old would be added marks against us.

❧

BACK IN THE FIFTIES, MIXED MARRIAGES WERE RARE, AND really frowned upon—especially to our recent enemy, the Japanese. I know that I've made mention of my occasional "winces" but, trust me, I'm downplaying the whole situation.

It's hard to imagine, since mixed marriages are considered chic today. I'm reminded of a lady about my age whom I ran into recently at Bloomingdale's. She was accompanied by her obviously Japanese, and beautiful, daughter-in-law and her two small "half-breed" grandtots. When I remarked on the children's beauty, she demurred pridefully:

"Yes, they're half Japanese."

I really demurred. I didn't let on.

❧

AFTER ABOUT A MONTH, WE HAD TO EXTEND OUR SEARCH ALL the way to Bailey's Crossroads, Virginia (about an hour's drive outside of Washington). Joe dutifully headed into our first target. When he came out, I was amazed. He was actually smiling. He reached into the car, grabbed my hand and said, "Come on, Bina, they're going to meet my family!"

It turned out that the rental agent had a brother who'd served as an officer in the 442nd. We finally had our apartment.

It seemed we had just settled in when it was time for yet another search. This time for a job.

⊱

THE COLD WAR HAD CHANGED THE RULES. IN FACT, THERE were no rules. These were frightening times. The Soviet Union, our former friend, was turning into a formidable foe. Everything about these guys was creepy—from their abysmal tailoring to their haircuts. And they were lethal. Their no-frills objective: world domination, without firing a single shot. Their means: fomenting revolution. Their method: use of their spy agency, the KGB. Bribery, blackmail, sabotage, propaganda, and even murder were not off limits. The United States had to do something. Our country needed its own spy agency.

The Central Intelligence Agency was created by Congress in the summer of 1947. It was tough going at first. There was bureaucratic infighting among senior staff members at the State Department, who feared the Agency would trample on their diplomatic turf like some rogue elephant. The Departments of War and Navy weren't too keen either; in true Washington fashion, they wanted to protect their own domain. A new agency that reported directly to the president might threaten *their* intelligence empires.

Critics charged that the wartime precursor of the CIA, the Office of Strategic Services (OSS), was just trying to perpetuate itself under a new name. And they were right. Many OSS staffers just didn't want to return to their boring Wall Street brokerage houses or their staid Washington law firms. Their war had been pretty exciting.

Joe shared their sentiments. He wanted to be on the front line in this new war; he saw it as a natural, and important, extension of his World War II infantry experience. It wasn't just that he craved excitement, he craved a worthy cause (and adversary). And he'd found it.

Before he graduated from SAIS, some recruiters from the Agency met with Joe on campus. They'd shown some interest in him, but their lengthy security check had us worried.

❧

SHORTLY AFTER PEARL HARBOR, WHEN JOE WAS TEACHING
on *Lanai*, he would use his shortwave radio in the evenings to get
news accounts of the war. First, he'd tune into the American reports
out of San Francisco. Then he'd flip over to the Japanese-language
broadcast out of Tokyo. After the Battle of the Coral Sea, Joe heard
U.S. accounts of our victory, at the same time that the Japanese
proclaimed a "stunning defeat of the American cowards."

The next day, over coffee with some "friends," Joe openly questioned what the real outcome of the battle had been. Within days,
the FBI was at his door, brandishing badges. The agents wanted to
know where Joe had gotten his information. Joe explained that he
had listened to the Japanese broadcast. They seized his radio and
abruptly informed him that the U.S. had won the battle. The FBI
tailed Joe thereafter.

❧

JOE'S FIRST JOB OFFER CAME IN—IT WAS FROM THE STATE
Department. They wanted him to go to the Micronesian island
of Nouméa, then a Trust territory, to add a little "color" to the
staff. We looked up Nouméa in an encyclopedia and learned its
biggest claim to fame seemed to be a high incidence of elephantia-
sis. Joe passed. We waited, watching our savings and Joe's G.I.
Bill benefits dwindle.

July 25, 1949 was our second anniversary. Even though funds
were short, I was determined to make it a celebration. Our daughter, Mary, was a year old, and we had another baby due in a few
months. I borrowed some money from my mother and planned
a dinner party for three.

Lights were dimmed, candles lit. In the center of the drop-
leaf walnut table sat our prized wedding-gift crystal canister set
entwined with hand-cut petunias (filched from the neighbors).
Mary was perched in her highchair, proudly patting the ruffles on
her electric blue and scarlet muumuu, a birthday gift straight from

Molokai. (I think she thought the party was for her.) A dinner of steaks, baked potatoes and frozen peas was served. Joe had just finished saying grace and was pouring the champagne when we were interrupted by a knock at the door.

It was Daddy, beaming.

"The CIA always gets their man. Congratulations, Joe."

Because we couldn't afford a phone, Joe had given the Agency my parents' phone as his contact number. Joe was to report to room 360 of the Statler Hotel the very next day. He was in. Now that was a celebration!

Of course, I didn't really know what we were toasting, except the prospect of a paycheck. I knew nothing of the CIA—except that it was the job that Joe wanted. And that was enough for me.

✥

THE AGENCY WAS ONE BIG ADVENTURE WHEN JOE CAME aboard in the late forties. It was run by some holdovers from OSS—Frank Wisner, Tracy Barnes, Dick Helms and Bill Colby—and manned by a bunch of amateurs. Most of them were Ivy League schooled, and many came from good families—some with money as well as position. They were an elite group—intellectual, imaginative and enthusiastic (good-looking, I might add). Man for man (and, while there were few back then, woman for woman), they epitomized the highest caliber of government servant.

Theirs was an exclusive club. Acceptance hinged mainly on ability. CIA attracted its share of mavericks, Joe among them, but they all shared one common denominator—patriotism.

Unlike many of his colleagues, Joe had no great family to back him up; he had no Ivy League education to open doors; he had never belonged to a club. It was at the Agency that he found his club, his niche, his friends—and his calling. The CIA offered him something that no other government agency could—anonymity—and a chance to kick over the traces.

Joe was a natural. He'd left Hawaii (and its persistent rumors) behind and was ready to reinvent himself. A spy? Perfect.

❧

JOE'S FIRST SESSION AT THE STATLER WAS A REAL DOWNER. He met with a group of ex-FBI agents, who were supposed to brief him. They immediately swore him to secrecy. He was to tell no one who he worked for, what he did, or with whom he had met. Their commands were a cinch to follow. He knew who he worked for, all right, but had no idea who he had just met with, or what his job was.

Joe was given the cryptonym Benjamin F. Okin, and promptly suggested "Franklin" for a middle name. No one cracked a smile. Then he was given his first assignment. It sounded pretty vague. He was to do research on Soviet life and call in weekly. (I can just picture Joe at the Falls Church public library in a tug of war with some seventh grader over the "S" volume of the Encyclopedia Britannica.)

When he came home that night, I expected to see an exuberant Joe. Instead, I faced a paranoid husband.

"Joe, tell me what happened?"

He escorted one-year-old Mary from the room, closed the door, put his finger to his lips, and in a hushed, deliberate tone, swore me to absolute secrecy. He then proceeded to tell me absolutely nothing.

The same scene must have been played out in other Washington homes. A couple of weeks later, we were invited to an office party given by one of Joe's new colleagues. My marching orders from Joe were clear. I was to reveal as little about myself as possible. (I was six months pregnant; that much they would know.) Joe was very definite; I was to watch my every word and refer to him only as "Ben." This was supposed to be a party?

Off we went, the merry couple—Mr. and Mrs. Benjamin Okin. Joe drove as I sat silently by his side. Saying little and listening more was a whole new role for me. I was practicing. Once, I called him "Joe" in the car. His response was swift: I was to get in the habit of calling him "Ben" for as long as the training period lasted.

We arrived at the couple's Virginia rambler. As I warily exchanged first names (our real ones) with some of the wives, it became clear that they had received similar instructions. None of us dared have a drink. We were really overdoing it and couldn't help glaring at our husbands, who, drinks in hand, were whooping it up in the other corner.

❧

THE RECRUITS WERE PROBABLY LOOSENING UP BEFORE THEIR next major hurdle: the following Monday, they were to get their assignments.

CIA employees are divided into two categories—overt and covert. Most CIA employees are overt and work in the area of intelligence analysis, administration, or technology; they can openly state that they work for CIA. Covert operatives cannot.

Covert operations accounted for less than 15 percent of the Agency's personnel, and its members were handpicked. Many of Joe's fellow recruits would be assigned to the espionage (intelligence gathering) and counterintelligence side of the Agency known as OSO (Office of Special Operations).

By the time Joe had joined the Agency, there was a new kid on the covert block called the Office of Policy Coordination (OPC). OPC's main focus was on influencing governments overseas, often through propaganda and psychological warfare. It wasn't enough to know what the Soviets were up to; we had to do something about it.

OPC had gotten its start in a national security memorandum drafted in 1948 by George Kennan, the leading Soviet expert of the day. The actual marching orders for OPC, which are now part of the public record, were to counter the "vicious covert activities of the USSR, its satellite countries and Communist groups to discredit the aims and activities of the U.S. and other Western powers." The operatives would have at their disposal the so-called black arts of espionage: propaganda, economic warfare, psychological warfare, sabotage, demolition and assistance to resistance groups. They were to carry out their activities in such a way that if ever revealed, the American government could "plausibly disclaim any responsibility."

The OPCers were the cowboys, the rabble-rousers, the coupmakers. To me, they were the cavalry of the Agency. They prided themselves on their ability to recruit, and control, agents—to operate. (As Joe later explained to me, contrary to popular parlance, covert Agency employees are called "operatives," not "agents." It's the foreign nationals who are recruited to work with the operatives who are called "agents.") Years later, journalists would give this branch of CIA the colorful monikers of "skunk works" or "dirty tricks" branch.

Monday morning arrived. The recruits were assembled in a large conference room. Their names were called out, followed by their assignments. Each recruit was then supposed to stand and depart through a designated door. The announcement came:

"Benjamin Okin, OPC."

❧

OSS. CIA. OSO. OPC. Agents. Operatives. Cold War. It was all new to me; it was new to everyone. I had plenty of questions. Where would we be stationed? (I wouldn't have minded going back to Latin America.) Would I ever know what my husband was doing each day, much less whether he was in danger? What would be my role? Could we ever tell anyone what Joe's job was? How would our kids be affected? Would we ever be able to speak freely in our own home? And when could we cut out this "Ben" nonsense?

But I was not to have answers. Not yet. Life would start to unfold on a need-to-know basis. I was to be patient and to have faith in Joe.

❧

THE TRAINING FOR CASE OFFICERS, AS OPERATIVES ARE KNOWN within the agency, lasted six months. It included an overview of surveillance, disguises, bugging techniques, defensive driving, marksmanship, clandestine communications, and sabotage. Joe learned how to use signals to make contact with an agent—a folded newspaper held at a certain angle, a cigarette handled a special way. Danger signals were taught. If a contact spotted hostile elements, the contact would change the way he was told to hold the newspaper or cigarette.

At "The Farm," a CIA training facility near Williamsburg, Virginia, things got pretty physical. The recruits learned how to pursue, evade and ram cars. They also learned marksmanship: they were placed in darkened rooms and instructed on how to point-aim at lighted objects. American and foreign firearms were used.

The most important training was in psychological warfare and covert action. Recruits learned how to gain access to and influence key government leaders; techniques for placing favorable editorials or articles in newspapers; even how to corner a commodity in overseas markets and other forms of economic warfare.

Joe's training hit a snag when it came to disguises. He called one night to tell me not to hold dinner. He'd be late. He was— about six hours.

The same thing happened the following evening. I waited up. Joe seemed discouraged, defeated, in fact.

He told me a few days later that a whole team of disguise artists had been working on him using mustaches, goatees, wigs, glasses, fake noses. He was to meet a Latin agent at the airport "incognito." Nothing worked. No matter what they did, he looked like a 6'4" Japanese man wearing a disguise.

They finally just gave up, rationalizing that no one would take Joe for a spy. The University of Hawaii newspaper had been right: Joe was just too noticeable.

❧

BUT IT'S NOT JUST THE TRICKS OF THE TRADE THAT MAKE A SPY. It's the ability to handle people. The most important people to handle were agents, who were often risking their lives to provide information.

Selecting and recruiting an agent was crucial. Prior to actual recruitment, a potential agent had to be carefully screened. Family background and any prior arrest record were especially important. Detailed psychological evaluations were done to uncover hidden influences and vulnerabilities. Was there anything in an agent's past to make him subject to coercion?

Agents were often motivated by money, or the sincere belief that collaboration would further the aims of both governments involved. But usually money.

Once an agent was recruited, control was paramount and ongoing. Their information had to be corroborated, their bona fides checked, their cooperation concealed. After all, one could be, and sometimes was, dealing with a double agent.

I think that, ultimately, what makes an exceptional spy is the ability to inspire trust. And people—be it agents or coworkers—trusted Joe.

⊱

WHILE JOE WAS IN WASHINGTON, AN UNMARRIED WOMAN IN his office confided in him that she was pregnant by a married man. Frightened for her job and herself, she was considering an abortion.

Joe invited her to our house for dinner. We tried to convince her to have the baby. Joe made it clear that if she had the baby, he'd watch out for her at the Agency.

I walked her back into Mary's room, and pulled out the small layette I had been assembling for our second child, who was due any day now. I took a few outfits and put them in a box for her.

Years later (twenty-three to be exact), Joe received a letter from this woman. Her son had just graduated from college. She had never married; her son was everything to her. The letter said that she never would have had her son if it hadn't been for Joe and that dinner.

❧

JOE WAS RAMMING CARS AT "THE FARM," I WAS SORTING laundry. I didn't mind—I was learning to be a mother and home-maker. It was just kind of quiet.

And I mean quiet. We had no radio. The rest of the country was listening to Nat "King" Cole or watching Bob Hope and Dorothy Lamour in their "road movies." Not the Kiyonagas. We had no car and darn little spending money. I was wearing out the eight classical records I had bought secondhand after taking a music-appreciation course at Michigan.

Television had arrived in many living rooms, but not ours. Our upstairs neighbor had a set and would invite us up on occasion.

I made it a point each night to change my clothes, freshen up, and dress up Mary to welcome her father home. It meant a lot to Joe and me. In some ways, we were a picture-perfect family.

❧

IT WAS SUNDAY. MARY, ABOUT ONE YEAR OLD, WAS SEATED in her high chair, the morning sun dancing off of her platinum-blond curls. Joe went to get his camera, a Leica he'd picked up on the black market. (I hated it every time I saw Joe with that damn camera. It took him so long to adjust whatever it was he adjusted that by the time he was ready, faces were long, hearts were heavy and the moment had passed.)

He focused on Mary. She looked up at him, her blue eyes smiling.

Then suddenly, Joe stopped. He wasn't smiling.

"Mary is not my child."

"What?"

"She's not my child."

"What do you mean, Joe? She's you all over again."

"Where did the blond hair come from? Neither of us has blond hair."

With that, Joe threw down the camera, stomped into the bedroom and slammed the door.

Talk about pictures. Suddenly everything snapped into focus. All the comments in Hawaii suddenly made sense. All of the problems with Joe's mother, all of the touchiness about his past. I suddenly realized what I'd gotten myself into.

Joe wasn't sure who his father was. He'd always been on unsure ground. He'd looked to me to solve his problems. He'd been looking to me—and our children—to reaffirm his Japaneseness.

Mary didn't look Japanese enough. His mother hadn't failed—maybe she wasn't pure as the driven snow—but I'd failed. I'd produced a blue-eyed, blond beauty, but that was not enough. I hadn't reaffirmed his mother's status as an "upright" woman; I hadn't established Junzo's position as his rightful father. I hadn't, beyond doubt, established Joe's identity.

Tough! So he wanted to shove his burden off onto my shoulders. He could try but it would never work.

Not as long as I was as outraged as I was at that moment. Not as long as I was as hurt as I was at that moment, and most certainly not as long as I was as terrified as I was at that moment.

I'd faced my eternity of winces, but now I was facing something far worse—a real possibility of an eternity of loathing.

We managed to sit down following my epiphany and have a head-to-head discussion. We had it out. Joe poured forth doubts, sadnesses, and profound love for his mother. Her many victories over daily tragedies—for his sake. Their privations and stoicisms together were finally brought home to me. I understood, somewhat.

And our lives went on . . .

❧

DAVID KIYONAGA WAS BORN ON JANUARY 11, 1950. I'LL BE honest: he was our homeliest child, somewhat resembling a pint-sized Buddha, though he has developed into a handsome facsimile of his father.

Our children never showed any evidence of sibling rivalry.

They were happy to have company. What's more, now Mary had someone she could boss around.

Mary and David first attended Sunday Mass when they were three and two years old. Usually, Joe and I attended Mass separately, rather than submit the rest of the congregation to possible upset. One Easter Sunday was an exception. Mary and David wore matching gray-flannel coats. (I bought most of their clothes at Best & Co.'s boys department since the tailoring was good and served for either sex.) They were visibly awed by their first Mass, the church, and especially Father McCarthy. He was a mountain of a man weighing close to 300 pounds. As we left Mass, he stood by the door and greeted each parishioner in turn. As Mary approached, she took his hand and whispered,

"Hello, God."

JOE WAS ANXIOUS FOR HIS FIRST OVERSEAS ASSIGNMENT. BUT his initial orders weren't exactly what he'd expected.

"Ben," his boss said, "you're in luck. You're about to embark on a cruise to the Greek islands."

Joe was immediately suspicious. Greece was hardly his province. Japan would have made more sense.

"You're to be a cook on a CIA-owned freighter bound for Greece. Once there, you're to jump ship and make your way through the mountains."

"Go on."

"Well, you're to cover about one hundred miles of terrain and make contact with Greek guerrillas."

"And then?"

"Once contact has been established, you're to train and lead the guerrillas as counterinsurgents."

Joe was stupefied. "You can't be serious. I can't speak Greek. I can't cook. And I'll never fit inside a galley."

His superior was incredulous.

"What's the matter, you lack guts?"

"Maybe," Joe replied, "but I've got a wife and two children, and I'm not about to risk my neck on a GS-7 salary."

The Agency relented. He was assigned to the Japan desk, and given back his rightful name (thank God).

Joe was to know his parents' homeland as a spy.

CHAPTER VI

❧

RED SAILS IN THE SUNSET

*I, Joseph Yoshio Kiyonaga, do hereby affirm my allegiance
to the United States of America and, likewise, disclaim any
partiality toward the Japanese by virtue of having blood relatives living there or by the fact that I was a dual Japanese/
American citizen until I was twenty-one years old.*
          June 1952

❧

OR WORDS TO THAT EFFECT. I NEVER ACTUALLY SAW THE OATH
CIA wanted Joe to sign; Joe never brought anything home from
the office. He mentioned it to me only in passing. I had my usual
reaction. I winced.

Here you had Joe, who grew up with everyone questioning
who he was, got tailed by the FBI for listening to the radio, and
then volunteered to slog his way across Italy and France fighting
the Nazis. He and his 442nd buddies had proved their loyalty
with blood, not signatures. Now he was being asked to prove

himself all over again. Joe had no problem pledging his allegiance; he just didn't like being forced to. I wondered if anyone was ever going to cut my husband a break.

Maybe it was just a product of the times. Oaths were big in the McCarthy era. Or maybe the Agency was just being especially cautious—Joe was one of the first Nisei operatives posted to Japan.

A stroke of the pen stood between him and where the action was. Joe wasn't about to sit out the Korean War behind a desk at headquarters. He was an operative. He wanted to operate.

Joe signed.

꙰

A JOURNEY OF 10,000 MILES BEGINS WITH A SINGLE STEP—IN our case, toward a Yellow Cab. In September 1952, the Kiyonagas said goodbye to their dinky suburban Washington apartment and, at dawn, piled into a taxi to National Airport. From there, it was on to San Francisco. We only made it as far as Chicago, where we were grounded by an airline strike. The rest of the way would have to be by train.

Now *that* was a train. The starched white damask tablecloths; the real rosebud in the silver vase; the crisp celery on ice. The chairs in the bar car were plush burgundy velour and, as Mary and David fast discovered, could swivel. At night, the porter pulled down the sleeper bunks in our compartment. We'd drift off to the gentle tug of the tracks. Things have changed. Back then, half the fun of traveling was getting there.

And back then, people dressed to travel. Joe, the proud father, would walk into the dining car each night with four-year-old Mary, who, in turn, would lead in two-and-a-half-year-old David by the hand. I dressed them in navy pullovers and matching Black Watch-plaid skirt and pants.

I think it was on that train that Mary had instilled in her a sense of taste and place. She was elaborately polite to the attentive waiters. She just didn't eat dinner, she dined. David refused the waiter's kind offer of a high chair. No sir, no high chair for him. He was a traveling man now, though he did permit me to tuck the large dinner napkin into the collar of his broadcloth shirt that peeked out from under his sweater. We were all amazed to see how the waiters managed to gracefully balance the trays of food as the train careened full tilt through the mountain passes.

The double-decker observation car, the "Vistadome", afforded leisurely and spectacular views of the jagged mountains and post-card-perfect forests. I'd flown over the Andes, but nothing compares to the Rockies. Later, in Japan, I would think back to how vast and green my own country had seemed.

The last time Joe had been on a train was when he'd gone cross-country to Camp Shelby. Now he—no, we—were heading in the opposite direction to the front lines of another type of war, one that was being fought secretly across the globe, with spies, agents, eavesdropping devices and disinformation campaigns as the weapons of choice.

The Soviets now had the Bomb, thanks to some adept spying on their part. British diplomats Burgess and MacLean had escaped to Moscow after spying for the Russians in Washington where they'd worked as British diplomats. (Their comrade, Kim Philby, the liaison between British Intelligence and CIA, would follow them into exile in 1963.) Alger Hiss, the State Department's fair-haired boy, and also an alleged Soviet spy, had been convicted of perjury.

After putting the kids to sleep, Joe and I would settle down in the bar car and he would light up a Chesterfield. (Or whatever brand it was that they had given out for free during the war.) We'd share the newspaper, silently passing the sections back and forth between us. The headlines screamed out at us. U.S. PLANES BOMB HYDROELECTRIC PLANTS IN NORTH KOREA. CHINA REJECTS U.N. ARMISTICE PLAN. 16,000 FLEE EAST BERLIN TO WEST. I listened. The world was getting personal.

Striking must have been in vogue. We arrived in San Francisco only to learn that the stevedores had walked off the job. What was to have been an overnight stay with a war buddy of Joe's—

Lindley Sale and his wife, Peg—became four days. We took full advantage, sampling eveything from the dim sum on Chinatown's Grant Street to the old-fashioneds at the Top of the Mark (the site of many a celebration for returned veterans).

We finally boarded our ship to Japan, the USS *Cleveland*, and were steaming toward the open sea for our fourteen-day journey. I was in our room dressing David for dinner when we approached the Golden Gate Bridge. The fog was so dense that you could only see the outline of the bridge. Joe was on deck, Mary's hand clasped in his, as the ship's foghorn sounded. At that moment, as if to comply with the horn's request, the fog lifted, providing a perfect view of the bridge. Later Joe told me that he felt for an instant as though the bridge were saluting the ship and that he should return the gesture.

It wasn't a luxury liner, but the *Cleveland* was huge (it seemed at least the length of a football field) and plied the Pacific with cargo and about a hundred passengers. The crew members were suitably craggy and flinty-eyed and seemed to be constantly burnishing the brass fittings and swabbing the decks of Balinese teak. It was to water what the Pullman had been to land: comfortable, real, and made to last—a far cry from the strangely antiseptic cruise ships of today that go in circles for entertainment. (And an even farther cry from the sampan from Molokai to Maui.) The *Cleveland* was Old World transportation. No, more than that—a transition, at least for us. And we savored it.

If I hadn't already been in love, I would surely have fallen in love with Joe on that ship. There are certain times in your life when the very air seems different—I think of our childrens' weddings or dinner parties with mariachi bands. Add to that the *Cleveland*. The limes seemed limier. The wind windier. The dinner table banter more natural. The search for your cabin keys more urgent.

The combination of the sun and vibrant blue Pacific by day contrasted with the stars and vast ocean by night. It was mesmerizing. So mesmerizing that I recall staring—and staring—at the immensity of the ocean one evening. It carried me along. Alone at the ship's rail, I sipped the last of my champagne, and, without thinking, flung my glass overboard. I watched, in fascination, the shimmer of the sheer crystal tumbling in the ship's wake.

As if to complement the blazing sky and sea, there was morning bouillon on deck served with a sliver of lemon; afternoon tea (you were offered a choice of dozens of exotic-sounding "chas" brewed on a silver tea cart accompanied by warm scones slathered with Indonesian honey); wonderful meals (the flaming desserts were almost too much); and, best of all, a full-time nursery.

Or maybe best of all were the *Cleveland*'s passengers. Some were right out of a Somerset Maugham novel: the quiet Jesuit cleric bound for Macao; the telephone executive en route to Hong Kong; the Ivy League historian visiting Jakarta; the Foreign Service and military couples. Joe might have been the only CIA operative on board, but I can't be certain.

Usually, Agency operatives overseas had "cover" job titles to mask their true trade. There are two types of covers: official (State Department, military, etc.) and nonofficial or "deep cover" (private industry, academics, etc.) The farther removed a deep-cover agent is from headquarters and official channels, the deeper his cover. Whatever his cover, a covert operative leads a double life and so does his wife.

Often, the operative is given an innocuous-sounding government position, the more innocuous-sounding the better. The goal is for the cover to be so boring or so vague that, at a cocktail party, when you mention that you work at, say, "the Government Accounting Office," your fellow partygoer's eyes immediately start to graze the room for more interesting company. In Joe's case in Japan, his humble cover was a "civilian with the Department of the Army."

My husband had looked great in the brown tweeds and dove-gray cashmeres of a Washington autumn. But now he was even more in his element, with a casual reddish tan to match. For the first time in months, he looked relaxed. And on the verge of so much.

❧

THIS SECOND HONEYMOON OF SORTS GAVE ME A CHANCE TO
think about our impending adventure, and about someone we'd
had to leave behind.

I'd become pregnant one year earlier. Twice during the pregnancy, I'd had to be rushed to the hospital, threatening to miscarry. The doctors found that the placenta was misplaced; the
medical term is "placenta previa." One doctor even went so far
as to suggest that it would be just as well if I did miscarry since
my hemorrhaging could cause brain damage to the unborn child.
I waved away his concerns, certain that the baby was sound. *I*
was the problem, and all of my problems stemmed from Mary's
horrific birth. I was determined that it would work out.

It didn't. The baby, a doll-like baby girl, was born two and one-
half months early. She was in an incubator when she contracted
a respiratory infection and died. She was three days old.

I had been confined to my bed and had not been allowed to
see her. We never named her. I understand she had red hair.

❧

MOST DAYS WE SPENT IN OUR ASSIGNED DECK CHAIRS IN THE
still heat, soundless except for the occasional splash of vermouth.
It put me in mind of Thanksgiving mornings when I was little
and I'd stay half asleep in bed. I'd smell someone roasting a turkey
and hear snippets of faraway conversations and the clattering of
dishes in our Baltimore kitchen. This had that same sense—life
beckoned but it was also strangely distant. And it could wait.

Which was fine for Joe and me. But imagine what it must
have been like for a four-year-old to be on the *Cleveland* in the
middle of the Pacific. To look out at nothing but water for days
as the prow of this gargantuan ship lunged toward our new home,
"Japan?" The kids, naturally, had questions. What did the Japanese eat, Mary asked? "Raw fish" is always an entertaining answer
to a four-year-old. David wanted to know if we could get him a

dog in Japan. I steered clear of that one. I was under the impression that the Japanese ate dogs.

I knew next to nothing of Japan—my mental images tended toward tintypes of Fuji, simpering geishas behind paper fans and scenes from Gilbert and Sullivan's *The Mikado*. Japanese history and culture weren't big subjects in Baltimore Catholic schools in the 1930s. More recent images included dive-bombing kamikazes, the USS *Arizona* enshrouded in smoke, and the five Sullivan brothers, all killed during the war while serving on the same ship in the Pacific—none of which made for great bedtime stories.

I had had my own questions for Joe. Would we need to bring a supply of vitamins? How about my mother's silver tea set? What about clothes? Would anything over there fit my five-foot, ten-inch frame? Thanks to the Agency's rigorous indoctrination, Joe may have been the expert on things Japanese—the politics, culture and sociology—but his knowledge didn't extend to the practical. Our whole life was in a permanent "wait-and-see" mode.

Five days out of San Francisco, we approached our first port of call, Honolulu. What was it about me and that place? Hawaii was like some sort of homing beacon—even in the vast Pacific we had to stop by to pay our respects to her. I had a vague uneasiness as we cruised by Diamond Head in the early morning. I wish we'd just sailed right on past.

❧

WE DOCKED IN HONOLULU AT 7:30 A.M. ALL I SAW ON THE pier was Joe's mother with two Japanese men—and as far as I could see, they weren't holding any leis.

Mother introduced us to her lawyers.

No mention was made of our baby's death. Instead, we were hustled off to the lawyers' offices where Joe was presented with some documents. We weren't even offered a cup of coffee. Joe was being asked to sign away his birthright—three rental properties willed to him by his late father. His mother felt she should have them because I was "not a good daughter-in-law, and would not take care of her in her old age." (That's a direct quote from the legal document.) She was right about that; I wasn't exactly in the mood to take care of her.

Joe and I were in shock. He consulted with me, while I waited with Mary and David in a side room. The rental income made a big difference on a lowly civil servant's salary. It was a cruel thing for Joe's mother to do.

I suggested that he sign. If she was willing to go to such lengths, let her have it, I counseled. He thought about it, and signed.

I looked on the bright side. She'd shown her true colors to Joe. To me, they'd always been about as subtle as an aloha shirt. She had cut her ties. From now on she would visit us less and dictate less. Joe would be more my husband and less his mother's son.

I expected that Joe, after this episode, would treat his mother with quiet disapproval, principled indignation . . . at least, a word or two about what she'd done. Instead, he was cordial, downright friendly, visiting with her in Japanese. Didn't he get it? Looking back I wonder why I didn't just head to the Pan Am terminal and back home.

But I didn't. I figured that if Joe could forgive his mother, I could forgive him for forgiving her. So I stayed and calculated the miles between Hawaii and Japan.

⊱

THANK GOD FOR THAT SHIP. EVERYTHING ABOUT IT WAS COM-forting after the stop in Hawaii. The passengers had become my new best friends—especially the Foreign Service wives—and I was loath to leave the warmth and security of their company. I felt like clinging to them as we made our way to Yokohama.

The trip had been leisurely. The Pacific had been so smooth—in fact, "pacific." I would have been happy to stay right where we were. Why not skip the whole Agency business and with it the prospect of an exciting life? Why not just skip Japan?

I couldn't handle any more surprises.

The gods must have sensed my mood and saw to it that our arrival in Japan was really miserable. About an hour outside of Yokohama, the seas turned rough. We all got seasick; the sky darkened to a threatening shroud of gray—and this was at eight in the morning. Then the horns hooted, chimes chimed. We had arrived.

My first glimpse of Yokohama was through our rain-streaked porthole as the *Cleveland* groaned to a halt at the dock. I've never seen such an unattractive town. It wasn't just war-torn; it was just plain ugly. Tan and dirty, it was an industrial port. Industrial and industrious—everyone on the dock seemed too preoccupied to notice we'd arrived. (At least Joe's mother wasn't on it.) Here I'd traveled across the world only to arrive at an Oriental version of Baltimore. If this was Japan, I wasn't going to like it.

But wait, maybe I would. I surveyed the hillside and glimpsed a grayish-brown house near the top. It looked like all of the other small, unpainted wooden houses on the hillside with one exception—there was a splash of crimson by its front door, a single flower in bloom! I was delighted. My first lesson in Japanese restraint.

Our driver met us at the dock to take us to our new home, about a forty-five-minute drive from the harbor. He was punctual, efficient, cordial, but hardly friendly. Midway through our trip, he abruptly stopped the car. He got out, walked to the side of the car and relieved himself. So much for restraint.

We reached our destination and new home: Kamakura, a coastal town about forty minutes southeast of Tokyo. Once Japan's ancient capital, it had become an artist's colony and, at first glance, I could see why. I felt as if we'd been plunked down in the middle of a vintage woodblock print. Japanese, in kimono, were everywhere. Fuji reigned from a distance and Buddha presided over all. And not just any Buddha, the famous *Daibutsu*. There he was, a huge and exquisitely crafted iron replica of the religion's founder, deep in a meditative trance.

I was entranced, but unsure. This was all so foreign. The kids were equally frightened. For once they didn't say a word. The unfamiliar can charm or alarm. Initially, Japan alarmed me; I wasn't sophisticated enough to be charmed.

We drove down the cobblestone road at Buddha's feet. Just short of the sea, we pulled into the driveway of our new home. Now this was more like it! It was an imposing, quasi-English manor house fronting the ocean with a meticulous garden of stone, wind-swept pines, and gentle mosses. My first introduction to Agency perks.

Standing by the front door was our second perk: our boiler man, Mori-san. His only job, it turned out, was to shovel coal into our home's boiler. Mori-san bowed. The Kiyonagas bowed back.

We entered our new house. It was fully furnished (Western style) with crisp linens on the beds and lots of fresh groceries in the icebox. The Agency was welcoming us—in style. I was surprised. It already felt comfortable, like home.

That is, until the next morning. I woke up late. No sign of Joe, just a note on the bureau. The driver had picked him up at 7:00 A.M.—he'd be home around 8:00 that evening.

Oh, great, just great! Here I was, on the edge of nowhere, with no idea where my husband was. No real idea where I was, stuck in a strange house with two small children—and Mori-san, who didn't speak a word of English.

Welcome to the CIA.

CHAPTER VII

❧

SAKURA

JOE HAD BEEN ASSIGNED AS CHIEF OF OPERATIONS AT THE CIA base in Atsugi, the U.S. Naval Air Station about forty miles southwest of Tokyo. He was part of what was known as JTAG (Joint Technical Advisory Group), the name given to the group that conducted covert operations out of Atsugi. (Talk about innocuous-sounding cover names.) The existence of the CIA base at Atsugi was a pretty well-kept secret.

Let me explain the basics, none of which I knew while we lived in Japan. Over the years I would learn that a base is a junior version of a station, the hub of CIA's operations in a given country. The chief of station, and his complement, or case officers, control the workings of CIA within the country, with the help of bases. The larger the country, the more bases. The station is located in the country's capital and is sometimes housed within the U.S. Embassy compound. Each station operates as an autonomous unit, but is always answerable to CIA headquarters. Any request from headquarters is given top priority.

Remember the U-2, the supersecret spy plane developed by the CIA? (I understand the name has since been co-opted by an Irish rock group.) The U-2 was an engineering marvel. It could fly at extremely high altitudes over enemy territory and take high-resolution photographs. (Word has it that it could photograph a grapefruit from 70,000 feet.) Getting information out of totalitarian regimes was very difficult back then, hence the critical importance of the U-2's surveillance capability. Some of the U-2s flew out of Atsugi.

No, I never saw one, but I'm pretty certain I heard one. A few years into our stay in Japan, we moved into temporary living quarters on Atsugi. Late at night, I would sometimes hear an eerie "whoosh." I had no idea at the time what it might have been. I only found out about the existence of the U-2 when the rest of the world did—in 1960, when the Soviets shot down a U-2 piloted by Gary Powers.

The CIA base at Atsugi wasn't big on decor—spartan in appearance, functional in approach, it was not a welcoming place. It was a large base, by CIA standards, encircled by a barbed-wire fence and consisting mostly of Quonset huts. There were dormitories on the base for unmarried personnel, and sixteen private homes were under construction for senior officers.

The base was located within the confines of the U.S. Naval Air Station. Armed Marine guards manned the entry gates at all hours. (The television comedian, Mark Russell, was one of the Marine guards while we were there; later, in the early sixties, Lee Harvey Oswald was also a Marine guard at Atsugi.) No one could enter or leave the base without "cleared" identification. CIA meant business.

And business, the spy business that is, was booming. Not just in Japan, all over—this was the heyday of the CIA's covert operations across the globe. The Agency's Radio Free Europe was broadcasting anticommunist propaganda behind the Iron Curtain. In 1953, the Agency successfully orchestrated a coup in Iran, handing the Shah the throne. The Far East was the other major theater.

Japan itself was of enormous importance, given its economic and political clout in Asia. The Japanese Communists were struggling, with some success, to gain influence, and it was CIA's job to thwart them. It was also up to the Agency to help ensure that a stable, democratic, and free-market government, loyal to U.S. interests, remained in power. (What the heck, we'd already written their constitution.)

Of course, the major event in the region at the time was the Korean War. Japan in the early 1950s was one of the principal staging and support areas for U.S. troops going to Korea.

We tend now to overlook the Korean War, but at that time, it was the testing ground for our resolve to contain the worldwide spread of Communism. By the time Joe and I reached Japan, the dreaded specter of Chinese involvement in the war was real; Chinese troops had streamed across the Yalu River in support of the North Korean Communists. The last thing anyone wanted was a protracted land war against a million Chinese. Intelligence on Chinese troop movements and war preparations was critical. Some of the most daring and innovative operations to gather intelligence on the Chinese Communists (or "Chicoms" as they were called) were based in Japan, as were many of the "psychological warfare" operations aimed at the enemy. It was a great place for Joe to start his Agency career.

Popular folklore would have it that every Agency assignment was cloak-and-dagger—black ties, martinis, poison-tipped pens, minicameras, the works. But it was usually much more prosaic than that. Take Joe's first assignment. The subject: a rusty underwater cable.

⌖

SOON AFTER JOE ARRIVED, TWO NATIONAL SECURITY AGENCY (NSA) staffers appeared at the office of the base chief, John M. Word of their mission had preceded them from Washington. They had a problem. Joe was summoned into John's office, along with Reese C., the base deputy.

The NSA, our nation's code-breakers, had found some success monitoring air traffic (radio messages) out of Peking (now Beijing). That is, until about a month before, when all air traffic stopped—right as the NSA was trying to break their code. It was suspected that the Chinese had switched to using an underwater cable to transmit messages. It was the job of the base to find the cable and cut it—no small order.

John, Reese, and Joe proceeded to tackle the problem. Which underwater cable? How would they locate it? Was it in hostile waters?

They knew of the existence of one submarine cable that spanned the Gulf of Chihli (a distance of some 250 miles) southeast of Peking. John recalled that the cable had been used for commercial purposes. Pinpointing it would be tough.

Reese suggested putting some Korean agents on it, but Joe had a better suggestion: "Jim Hawkins." Jim could go to Pusan and look into it firsthand.

You'd have to know Sgt. Jim Hawkins to appreciate just how unlikely a spy he appeared. Big, jovial, relaxed, complete with a beer belly, he seemed about as devious as Captain Kangaroo. But he knew his craft.

Jim was dispatched, but there were other questions to be settled. Once they found the cable, how could they surface it and cut it? The water in some places was thousands of feet deep.

Reese, an old Navy hand, proposed that they rig up a sea hook, drag it along the ocean floor and, once contact had been made, lift the cable to the surface and cut it.

Since no one else had any ideas, they decided to try Reese's. Reese and his team set about designing and building the sea hook.

Meantime, Joe and his group began calculating times, moon phases and tides.

❧

IF I'D KNOWN AT THE TIME WHAT JOE WAS UP TO, I WOULD'VE offered my considerable knowledge on the subject. Most of the day I spent alone with the kids sitting on the beach in front of our house (a beautiful beach, albeit contaminated), gauging the rising tide. The kids would wander along the beach collecting shells and odd bits of refuse, usually beer and sake bottles.

At dusk, we'd watch the fishermen in their blue and white

teneguis (headbands) haul in nets wriggling with squid and what looked like silver eels. Sometimes we'd be lucky enough to happen upon someone flying an artfully designed and brilliantly colored Japanese kite, a far cry from the newspaper kites that Daddy made for me back in Baltimore.

My fascination with the boundless sea was wearing on me, and I wasn't exactly comfortable at home yet, grandiose as it was. Mori-san and I were starting to give each other the creeps. I seemed to come upon him almost every time I'd round a corner. (He moved silently, as we didn't wear shoes in the house.) It was a little like sharing your home with someone else's cat.

After a few weeks, I decided to go into town with the kids. One morning the three of us began our descent down the half-mile dirt road that led to the town of Kamakura. I approached Kamakura that day—and the Japanese—with some trepidation. It was my first real contact with the country and its people. Everything about the Japanese intrigued me, but it was an intrigue mixed with distrust. I'd lost too many cousins, and my first boyfriend, Bill Linde, in the war. I wasn't in a real forgiving mood.

Without any money (it didn't occur to Joe to leave me any) and not speaking any Japanese, we ventured out into the maze of alleyways of Kamakura. The stores were colorless from the outside, with the proprietors in weathered kimonos to match. We were the only Westerners in sight.

We visited the closet-sized shops, each selling a particular item—homemade tofu, mounds of green and brown teas, delicate gold-leaf paper, earthenware pottery, hand-crafted knives. One shop sold iridescent animal toys—rose-colored cranes, aqua tigers—made of celluloid. The shopkeeper gave Mary a small paper crane as we left. You can't hold history against someone who is kind to your child. That small gesture did more to ease my distrust of the Japanese than any diplomatic initiative could have.

Eventually, we extended our daily walks past town, until we found ourselves face-to-face with our most famous neighbor, the Buddha. The kids and I began to visit him frequently—practically every day. Routine is the first step toward familiarity. Japan was beginning to feel like home.

❧

SGT. HAWKINS MADE HIS WAY TO KOREA SEVERAL TIMES AND managed to track down an old sea captain who claimed to know exactly where the cable was located. He also knew of a fairly shallow area, about fifty feet deep, where Reese's lift could be undertaken. The problem was that the area was very close to the Chinese shore and the ever-present menace of Chinese patrols and sweeping searchlights.

The mission was demanding. It required a crew ignorant of its task; a fast, small fishing boat, sufficiently large to house the cable-lifting mechanism; and a trusted skipper-agent who knew the waters. They had to cope with many hazards—icy water, swift currents, and working at night. Chief among their concerns was the time element. It was imperative that the ship enter and leave Chinese waters under cover of darkness.

The logistics were worked out. Reese fashioned a formidable sea hook, and Joe selected a propitious night and hour. Poor Jim was again dispatched to Korea, this time to round up a crew, outfit a boat and contact the skipper, a reliable agent.

Once recruited, the crew had to be initiated into the mysteries of wet suits, goggles, and the like. For what was to have been a simple fishing expedition, extraordinary training seemed to be in order. The crew grew restive and curious. Their pay was doubled, and the queries ceased. Numerous dry runs were undertaken. Maneuvering the sea hook required brute strength, skill, and repeated practice. The skipper alone was witting. Once on the high seas he was to advise the crew of what lay ahead.

D night arrived.

Early the next morning, a DC-3 flew into Atsugi bearing a very special cargo—a relieved Sgt. Hawkins cradling a salt-encrusted piece of cable. It measured two feet in length and six inches in diameter.

Joe was somewhat skeptical. What if "old reliable" Sgt.

Hawkins had just picked up a piece of rusty cable on the docks of Pusan? His suspicions were allayed when the NSA informed the base that the mission had proven an unqualified success.

The Chinese had resumed full air traffic.

※

AT LEAST THE CHINESE COULD COMMUNICATE BY AIR. I, on the other hand, barely saw, much less spoke, with my own husband. Joe was immersed in his work, out at 7:30 every morning and back in the evening right before the kids' bedtime at 8:00. In theory, I could use the phone, but for security reasons, I wasn't allowed to call him during the day.

Even when he got home at night, we couldn't really talk. What's the point of being married if you can't talk? There were signals, though. If Joe whistled as he walked in the door, I knew he'd had a good day. The bad part was not knowing what constituted a good day—a rusty underwater cable; a commendation from headquarters; or maybe, just a good cup of coffee. A bad day was easier to detect—no whistle, occasional drumming of his fingers on the dining room table (especially if dinner was late), a perceptible tensing of his jaw muscles.

I tell you, it was no fun, and I began to take it personally. How about me? I'd been sitting around all day with the kids, doing practically nothing. Couldn't I live vicariously just a little bit? It got to the point where I would have appreciated hearing even about his cup of coffee.

I needed someone, besides the kids and the Buddha, to talk to—even if it was just about the weather. I needed a friend.

※

WHILE SITTING ON THE BEACH ONE AFTERNOON, I SAW something—or someone—approaching. Through the shimmering heat and shifting sands, I spotted my salvation. I was amazed. There she was, and coming closer—an American. We both began talking at the same time.

"Hi, I'm 'Hope Turner' and this is my son, 'Tom.' "

She was attractive, pleasant, and our kids were about the same age. Within ten minutes, we were practically best friends.

Then began a series of exchanges about what our husbands did. "That's right, a civilian with the Army stationed at Atsugi." "Oh really, with the Navy at Chigasaki?" It went on—and on— as we elaborated on their phantom careers.

That night, Joe and I went to our first party for the CIA staff. The first person I ran into was Hope.

❧

I WAS BEGINNING TO GET IT. I WAS TO KNOW NOTHING ABOUT what Joe did. I was to tactfully deflect any inquiries about his job. When I did speak, I was to be careful about what I said, and the less said the better. Joe's philosophy was that what I didn't know, I couldn't repeat. Fine with me—it was for everyone's protection.

But the less you know, the more you suspect. The kindly shop-keeper inquiring in broken English about the size of our family might be helping compile a dossier. The neighborhood knife sharpener asking whether we would be returning to the U.S. for the New Year's holidays might be trying to determine whether the house would be empty. Could that one stray bit of information be the missing piece that the enemy needed to put together the puzzle of an Agency operation? You learn to be friendly, but vague.

When you miss someone, it can help to at least have a mental picture of where he spends his days. No chance—initially, security precluded me from seeing Joe's office. After a year in Japan, I was allowed to visit his office once, and very briefly. I'd expected to see some cloak-and-dagger paraphernalia—an electronic gadget or two, maybe a few technical maps. The office resembled a prison cell: a small window, spotless gray metal desk and file cabi-nets secured by steel bars and padlocks. Not a comfortable chair in sight. The CIA meant business, all right.

So in those early days, I had to conjure up my own mental pictures of what Joe might be up to. In the middle of my day, I'd find myself wondering where he was. Sipping scotch with who-knows-who discussing who-knows-what? Consorting with fawning geishas? Climbing Fuji? I couldn't be sure. All I knew for sure was that he wasn't telling me a thing.

Part of me thought that maybe the whole secrecy thing was just a big ego trip for Joe. (Was he really doing anything *that* important?) But I knew, from the daily newspaper headlines alone, that the stakes were high. And at least I finally had a role, even if it meant just keeping quiet. Problem was, I had no idea how to play it.

Japan itself came to the rescue. I took one look around and snuck a page from the Japanese wife's handbook. She was a study in diplomacy. She ruled the home, husband included, with an iron hand. But she did it in such a subtle, unassuming way that no one seemed to mind. A model of discretion, she became my role model. So a new Bina emerged—Bina, the subservient, uncomplaining Japanese-style wife. I went from being frustrated by Joe's refusal to tell me anything to being silently supportive and trusting. Joe was just plain puzzled. It seemed I'd beaten him at his own game.

Some times were more difficult than others.

※

THE EVENING HAD BEEN A FULL-BLOWN SUCCESS. WE'D invited some people over, and Joe had begun to foster a friendship with a government official. The kids were asleep. Tomi-chan, our housekeeper, had retired to her quarters. The ashtrays were full and plates of half-eaten sushi littered the tables—they could wait till morning. We had earned a cognac before going to sleep. Then the phone rang. Damn. (I still hate hearing a phone ring at night, even when it's a phone on a television show.)

"Joe, who was it?"

It was his mistress: the Agency.

"The office. Just checking in."

But at dawn, Joe kissed me good-bye. I glanced at him as he

headed out the door; something looked different. He was wearing military fatigues. He was out the door before I could ask questions.

It was a long day. At the end of it, no Joe.

I knew enough not to call the office—any tip-off of Joe's absence would be a breach of security. Would he be back that night? Where had he gone? Where in Japan would you need to dress in battle regalia? That day's *Nippon Times* provided a possible clue: U.S. troops in Korea were mounting a major counteroffense against the Chinese.

The kids had questions. I had no answers. (Actually, it took them a couple of days to notice he was gone.)

I marvel at the fact I didn't call someone—his boss, a friend. Anyone. But I didn't. Somehow I managed to keep my hand away from that phone. By the third day I was terrified to the point of being calm. Dead calm. I took my shower. Read some book to David, and another to Mary. Then, finally, the call came.

It was from Joe's office: I could expect him home within the hour. No explanations.

And then, in he walked. He was casual—unfazed—as if he'd just been out to pick up some dry cleaning.

"Joe, where . . ."

With that, he made a motion with his hand I would grow to hate: his four fingers clamped down on his thumb—"Not a word."

It was all I would hear about that disappearance until thirty years later. He'd been close to the action in Korea. Turns out Joe had overseen many operations related to Korea, including some employing psychological warfare. One involved beaming dispiriting radio broadcasts (à la "Tokyo Rose") to North Korean troops and dropping similarly themed pamphlets behind enemy lines. (The Chinese reciprocated by spreading rumors that the pamphlets contained a toxin—deadly to the touch.)

The Korean War ended in July 1953 and, with it, Joe's trips—for a while.

❧

WHEN YOUR HUSBAND LEAVES FOR DAYS AT A TIME WITHOUT anyone even calling you, you start to get the idea: you have to fend for yourself as a CIA wife. I wasn't mad at Joe; the circumstances were hardly his fault. So fend I did. My forced independence and solitude had its advantages—it gave me the chance to observe and eventually relish the evocative simplicity of Japan.

If you're going to know a foreign country, you might as well know it while it's still foreign. Everything about Japan was foreign: the cedary scent of polished wood and incense; the salty freshness of raw tuna spiked with *shoyu* (soy sauce) and daikon (grated horseradish); the clash of a spectacularly colorful kimono against a subtly hued garden of artfully arranged pebbles and rocks. I think my love of "shocking pink" originated in Japan. Their studied, and at first jarring, use of pink and scarlet in the same fabric and decor intrigued me.

Nature is paramount in Japan—it's what the Japanese have instead of God. Things I'd really never noticed before—the clean smell of pine; the trickle of running water; the soft touch of the spring *kaze* (breeze); the nuances of the changing seasons—took on new importance. In Baltimore, a change of seasons had only meant that I had to rummage through the attic for a change of outfits.

The Japanese minimalist approach to life extended to their homes. Their houses were small (even their castles were small), spare and immaculately clean. Of far more importance was the garden. Its every nuance reflected the personalities of the persons within—whether it be a bed of "paper-whites" protected by the fond shadow of a centuries-old pine tree; a venerable stone lantern with a single candle peeking out, surrounded by a mass of wild strawberries; or the extravagant severity of a rock garden, raked and watered each morning.

And I began to see the people differently. Underneath their reserved, refined, and cultivated exterior, I found that there beat an earthy—sometimes downright crude—heart. An extension of nature, I suppose. The nude body and attendant functions were dealt with matter-of-factly. (Except for blowing your nose, which I never once saw anyone do in public.)

And talk about intense—the Japanese approached everything, even the seemingly mundane, with ferocious efficiency. It's worth the trip to buy a gift at Mitsukoshi, a great Japanese department store, just to watch the frenetic choreography of the thing being wrapped. As for sushi, apparently it takes seven years to master the proper technique for hand-rolling them. (It's all in the wrist.) Even nature is whipped into shape: you can plunk down a whole month's wages for a single honeydew of amazingly precise contour and hue. Not to mention the beauty of hundred-year-old, hand-sculpted bonsai trees the size of sake bottles. I knew we were in trouble years later when those neat little Sony televisions started landing on our shores—how can you compete with a country where the putting greens are trimmed with manicure scissors?

Occasionally, they did go a little overboard.

※

JOE ENTERED THE DINING ROOM ONE MORNING AND RANG THE bell for breakfast. Tomi-chan appeared and stood sheepishly by the table, not even looking at Joe.

"I'll have some eggs, please. And coffee."

"*Moshiwake gozaimasen.* (I'm extremely sorry.) We have no eggs or coffee."

Joe wasn't too pleased.

"Well, just some toast."

"*Moshiwake gozaimasen.* We have no toast."

This wasn't looking good.

"Juice, then."

"*Moshiwake gozaimasen.*"

Only at that point did Tomi-chan get around to explaining that our pantry had been robbed overnight. Some poor, obviously hungry soul, had filched our entire supply of food.

The police were called and no less than the chief of police himself arrived. I couldn't take my eyes off his uniform—a blue serge masterpiece with epaulets and polished gold buttons. Then began the laborious investigation. Fingerprints were lifted, diagrams drawn, a complete inventory taken. Finally, the *coup de grâce:*

"Sir, were any other items stolen?"

(This to Joe, who had never set foot in the kitchen.)

I volunteered that there had been some saltine crackers in a box, but I wasn't sure if they'd been stolen.

The chief seized the initiative and disappeared into the kitchen for several minutes. He emerged triumphant, dangling a small—empty—cracker box. Eyes narrowed, he resumed his investigation.

"As of last night, how many crackers were there in this box?"

⊱

CRACKER THIEVES ASIDE, JAPAN BEGAN TO TAKE ON FOR ME AN almost mystical allure (or maybe I was just spending too much time alone). The Japanese have a word for things, or people, that have a muted, rough-hewn elegance—*shibui*. The term is reserved for those few things that don't call attention to themselves, but that you can't help but notice.

I didn't really have a sense of what the term meant, until I went shopping with Joe in Kamakura. We had stopped at a carpentry shop, which had some lovely pieces on display: teak chests or *tansus* and small Shinto shrines for the home. All were beautifully worked and polished, but what caught Joe's eye way in the back of the shop was not on display. It was an unvarnished wood piece, with large wrought-iron handles and a padlock. Solid and self-assured. *Shibui.*

"How much for that piece back there?"

The shop owner bowed graciously in obvious embarrassment.

"That unworthy piece is not for sale."

The shop owner was using the chest as a working safe, hence the padlock. He offered us a cup of tea while Joe opened the negotiations. Joe couldn't help but notice the spare, hand-wrought elegance of the ancient black iron tea kettle. We walked out the owners of both items. (Thank God the store owner wasn't wearing a halfway decent-looking watch.)

I came up with the idea of putting a white marble top on the chest—Bina's West meets Joe's East.

But the most striking example of the blending of Asian and Western came in the form of a Dominican chapel nestled high in the hills above Kamakura. I recall our first Sunday Mass. A waft of incense greeted us as we climbed the three wooden stairs to the open door. The chapel was Japanese style, with *shoji*, paper sliding doors, and *tatami* mat floor. The handcrafted wooden altar was watched over by a huge, unadorned cross of gnarled cypress. There were a few wooden benches and no windows—just greenery and sky all around—an extension of the Japanese idea of bringing the outside in.

On our first visit, the priest was American; the congregation (seated on the tatami matting) Japanese, except for one obviously American couple and son seated on a bench. Everything, including the sermon, was in Japanese. But it wasn't foreign in the least. Everywhere I've ever gone—no matter how remote or how alien—I've always had a place where I felt comfortable, a place I could always call home: the Church.

After Mass, Joe and I, Mary and David waited to speak to the priest and the American couple. It was then that I began to understand the concept of six degrees of separation. The couple were Navy; he was a commander at Yokosuka, the U.S. Navy base nearby. They were very fair—he and the son even had reddish hair and freckles. That should have given me a clue. Amazingly enough, he was my third cousin on Mother's side and had been chief of police of Wilkes Barre, Pennsylvania (or Scranton, I forget which). This couple became our good friends. It was through them that we learned of the rebuilding and rededication of a Catholic cathedral in Hiroshima that had been gutted by the atom blast, a seemingly small but astoundingly hopeful act that would later inspire John Hersey's work, *Hiroshima*.

❧

JOE AND I NEVER VISITED HIROSHIMA. BUT WE WERE IN *Japan in 1954 when the cathedral in Hiroshima was rededicated. The priests, all foreign-born Jesuits, had been there for years and spoke to each other . . . in Japanese. Some had survived the blast and were present for the rededication.*

When our son Paul visited Hiroshima in 1985, he found it was a bustling city with a surreal quality. As he was leaving, he was approached by an old, disfigured man at a train station restaurant. The man spoke loudly—and for the most part unintelligibly—in the most polite, honorific Japanese. He asked to sit with Paul and offered to buy him a beer.

The man began making an odd outline with his hands, as if describing a woman's figure. Eventually, Paul realized the man was a survivor of the bomb, and he was depicting the shape of a mushroom cloud.

Paul later told me how peculiar it seemed to have this victim of the bomb buying him, a young American, a drink. He thought it was the ultimate act of submission. Of course, Paul was young and didn't realize it was an act of forgiveness.

⁌

IT WAS ALSO THROUGH THE CHANCE MEETING WITH MY COUSIN in the church in Kamakura that we learned of Saint Mary's Orphanage in Yokohama.

After our infant daughter died, my gynecologists, eminent specialists, told us that it would be very "unwise" for me to get pregnant again. I had also suffered two miscarriages. We took them at their word. Their to-the-point prescription: adopt.

We headed for Saint Mary's Orphanage, home to about sixty Eurasian babies, mostly born of bar-girl mothers and G.I. fathers. Joe and I met with the Reverend Mother. We wanted to adopt two babies—one boy, one girl. Sister Lucille was most receptive. Joe and I tentatively selected two darling babies. (I hate to admit it, but it was kind of like picking out puppies.) We told Sister that we'd get back to her when we had organized a nursery. A

week later, I realized that I was pregnant and notified Sister Lucille. She was understanding and happy for me.

Unwise or not, I was pregnant—and hopeful. This time I dispensed with the advice of highly touted specialists and went to
see a young Navy doctor at Yokosuka. He was the only doctor I
saw until I went into labor. His counsel was simple: bed rest for
eight months. No stairs, no car, no walking around.

For eight months, I lay in our bedroom, staring at three white
walls. Why I didn't have sense enough to have our double bed
turned around and have it face the bank of windows and sliding
doors that looked over our lush garden and seascape is beyond me.

Nevertheless, there I was—flat on my back with little more
than a few records and some books to entertain me. (I've never
listened to *La Mer* since.) At one point, I even resorted to reading
a physics textbook. I was relaxed and didn't despair. I was on a
mission of my own now.

Despite his busy schedule, Joe was home each night by seven
during most of my pregnancy. Being confined to bed, my life
hung on telltale, familiar sounds. Joe's car as it rounded our gravel
driveway was my favorite.

He'd ready the kids for bed, and then carry them into our
room. I can picture Joe in his tan gabardine slacks, blue oxford
shirt, black knit tie, loosened, with Mary in one arm and David
in the other. It's hard to think of a time he looked better. Sometimes we'd sit together on our bed and read them fairy tales
with singularly Japanese titles like "The Foolish Jellyfish" and
"Momotaro, the Peach Boy." When storytime was over, Joe and
I would listen as the kids knelt by our bed to say their prayers.
Then he'd see them off to bed.

❧

SOME MONTHS EARLIER, HOPE TURNER HAD POINTED OUT TO
me that David had what's termed a "wandering eye." I call it
wall-eyes. When Hope called it to my attention, I took offense.

(It's much the same as seeing someone everyday; if they've gained weight, you don't even notice.) But she was right.

Surgery was scheduled at Yokosuka, which, fortunately, had world-class surgeons. Yokosuka served as a way station for the severely wounded from Korea.

Confined to my bed, I couldn't go. It was hard to watch Joe and David head off to the hospital together. After the two-hour surgery, David's eyes were bandaged. When he came to, he still had on the bandages. How do you explain to a three-year-old why he can't see, when he pleads: "Daddy, turn on the lights."

Joe stayed by his side until the bandages were removed the following afternoon.

❧

IT MUST HAVE BEEN REALLY TOUGH ON JOE. A FULL DAY AT the office, capped by a night of caring for his wife and children. He turned down many invitations to be home with me and the kids, invitations that no doubt were important for his career. He did it with great good grace and made me feel cherished in the process. Of course, he had to cherish me from a distance, so to speak.

After Joe'd tuck Mary and David in, our evening together would begin. Tomi-chan would arrive with our dinners on separate bamboo trays. Joe would sit in his blue velvet easy chair and visit with me while we ate. Later, Joe would give me my bed bath before we'd turn in for the night. Sometimes the least memorable times are the ones you remember most.

John Cady Kiyonaga was born early on a Sunday morning, December 27, 1953, after Joe and I had been out dancing. Since John was already overdue, and Joe saw the tax deadline approaching, he felt that a few turns around the floor could do me no harm. I went straight from the dance to the hospital at Camp Zama, a *M.A.S.H.*-like military installation between Yokohama and Atsugi. John made his appearance two hours later. If size at birth is any indication, John runs true to form. He was 22 inches long. He now stands 6'7".

We named him after my father, and interestingly enough, he resembles him in every particular—even to his wit.

※

JOE HAD STAYED HOME NIGHT AFTER NIGHT FOR EIGHT MONTHS to keep me company. But there's company and there's company. After John was born, I realized how long it'd been since we'd been out on the town. A dinner invitation from our neighbors, the Shimizus, came at the perfect time.

The next Saturday evening, Joe and I kissed Mary, David and John goodnight and headed out. I was excited, almost nervous, at the prospect of an evening out with my husband. A real date.

The Shimizus' house was smallish, surrounded by the ever-present wall and a classic Japanese garden, complete with a foot-bridge. They greeted us at the door—she in a muted kimono, he in a Western business suit. (He'd been educated at Oxford.) Joe and I removed our shoes and were proffered Thai silk slippers.

Happily, language was no problem. They both spoke English. Kazuko, the wife, had studied at the Sacred Heart School in Tokyo—the same school that the present Empress Michiko later attended.

Drinks were served. Scotch and water—no ice—British style. The host, Yoshio, took great pride in displaying the label on the bottle, OLD PARR. I'd never heard of it, but I later learned that it was highly prized in Japan. All foreign labels were prized in Japan. (So prized, in fact, that during the Occupation the Japanese tried to outlive their rinky-dink reputation as manufacturers by enterprisingly renaming an industrial town "Usa." All of the town's products, from toys to dinnerware, were laboriously stamped MADE IN USA. It sure beat MADE IN OCCUPIED JAPAN.)

Shoji doors were opened to reveal a beautifully set Japanese table. (On the floor, naturally.) Unlike our Western style of food service, the Japanese enjoy a delightful knack of mixing pottery, porcelain, and lacquer all at the same setting. I settled myself on a vermilion shantung cushion. "Settled" is hardly the word. I never, in my nearly six years in Japan, really figured out how to arrange my long legs in relation to the pillow and the tatami floor. For me, it was extremely uncomfortable. Next, the chopsticks. They were not the wooden variety that I'd run into in Chinese restaurants. These were ivory—and slippery. But not half as slippery as the first course.

There it sat, staring me in the face—a sliced tentacle of raw octopus, artistically arranged to resemble its natural coiled bent— all set to attack! After much good-humored (I guess) tittering by our hostess, I managed to grasp the uppermost slimy tentacle between my equally slimy (or so they seemed) chopsticks and convey it to my less than eager lips. Believe me, it wasn't worth the trip. It tasted like rubber—as in Goodyear. After much chewing, I managed to wash it down with massive quantities of sake.

The next course was sukiyaki. Now, *that* I could handle. As my hostess spooned out my portion from the bubbling cauldron, I began to relax. Either it was the sake or the equally intoxicating scent of the beef, mushrooms, leeks and *shoyu* wafting toward me. But no, just as she was about to place it before me she topped it off with a raw egg. What is it with the Japanese that they like so much of their food raw? There I sat having my second raw encounter in less than thirty minutes.

Next we were each presented with an exquisite gold lacquer bowl. As I removed the top, less than eager for my next adventure, I spotted gold-fish swimming in the crystal-like water.

My immediate thought: My God, these people really are heartless. They serve up tiny, innocent goldfish, blissfully swimming in their gold-encrusted pool and expect me to eat them, too! But I was beyond caring. I was just about to pick one up in my fingers (forget the chopsticks) and pop it in my mouth, college-boy style. Luckily, Joe motioned to me. What I was about to devour was a decoration—in our finger bowl.

Green tea and a *yokan* (soybean sweet) followed. Joe and I made our good-byes.

When we walked out into the midnight air we were surprised to find that it was snowing—pretty rare in that part of Japan. At that point even a typhoon wouldn't have surprised me. It was eerie as we headed home. No one else was about. No one else was enjoying the quiet spectacle of a snow-dusted Kamakura.

Joe stopped the car at the top of our cobblestone road. We were going to pay a visit to my old friend.

We left the car and approached the Buddha, our footfalls crunching as we walked. Scattered everywhere amidst the snow were fallen cherry blossoms, *sakura*. In the moonlight, my friend didn't look quite so friendly.

There he loomed, shrouded in silence. The consecrated hush of the freshly fallen snow embraced us. The chill crystal splendor of Buddha's size diminished us.

Joe grasped my hand; his steady strength reassured me. He was still there for me. He was still Joe. We were still us. Nothing had really changed.

The evening had been perfect—after all.

❧

CHAPTER VIII

TWO SPIES SENT OUT INTO THE COLD

JOE NEVER CELEBRATED HALLOWEEN AS A CHILD. WITH NO concept of trick-or-treating, no wonder disguises were such a fiasco for him.

We made up for those years after we got married. Halloween became a real event. When the kids were small, we'd all dress up alike—a roving band of "Frito Banditos" with holster straps was a favorite. The kids outgrew this ritual a lot quicker than Joe and I did.

Halloween was pretty high on the Agency's list, too. Nothing like a costume party to bring out the best in a room full of spies.

❧

IT WAS OCTOBER 31, 1952. JOE TOLD ME THAT THE BASE WAS throwing a big Halloween party. I went as TIME TO RE-TIRE (Firestone's slogan at the time), and Joe, as the "absentminded professor"—shirt and tie, no trousers.

There were some recurring themes among the costumes. "The Shadow" was a big one, as was Mata Hari. Even the base was appropriately disguised, filled with glowing pumpkins and draped with orange and black crepe paper.

I met two new recruits that night who made quite an impression on me: Jack Downey, a recent Yale graduate, and Dick Fecteau, of Boston University. Both had played on their respective football teams, and they looked it—athletic, rugged, just plain good-looking. They had energy, too, and that Halloween night they were having a ball. They made an impression on Joe, as well. And I don't think he ever forgave himself for it.

❧

ONE OF THE AGENCY'S MAJOR OPERATIONS OUT OF ATSUGI WAS aimed at supporting the Korean War effort. The operation consisted of training Nationalist Chinese agents at a large compound in Saipan and then dropping them by air in northern China behind enemy lines. The agents would then gather intelligence on Chinese troop movements and the flow of materiel into the Korean Peninsula. The planes used were part of the Civil Air Transport fleet, an airline the Agency had purchased from Gen. Claire Chennault of World War II "Flying Tigers" fame.

The operation was risky and proved difficult from the start. First, the agents needed proper paperwork. So the base took over a Japanese fishing boat, installed high-powered motors and armaments and skippered it with an ex-commander of the Japanese Imperial Navy. The boat cruised the Chinese coastal area, attacked small Chinese ships, and "obtained" the needed documentation. No survivors were left behind. Remember, this was war.

Assuming you got past the paperwork glitches, the air drops were still chancy at best. An agent was dropped with his radio and supplies at a selected spot somewhere inside China. He was to radio reports periodically and eventually exfiltrate himself. When he was all set, the agent would radio in that he was ready to be "skysnatched."

Skysnatching is more primitive than it sounds. It was developed at Atsugi and consisted of a manually operated grappling hook in the plane that lowered a harness made with bamboo poles. (It sounds like hooks were the CIA's secret weapon back in the fifties.) The plan was for the plane to fly slow and low—about sixty miles per hour—and scoop him up with the harness. Then they'd reel him up like a flounder and fly on home.

In November 1952, one of the Chinese Nationalist agents already behind enemy lines radioed in an urgent message—he had to be skysnatched immediately. He had run out of food.

A time and place were arranged for the "scoop." But on the afternoon of the proposed flight, two of the crewmen—the ones in charge of the grappling hook—called in sick. Right there, something didn't sound right.

John M. called an emergency meeting with Reese C. and Joe. It was their unfortunate duty to have to select two substitutes.

Reese suggested sending two Chinese Nationalists who worked for the Agency, but John nixed the idea:

"They haven't been cleared and I don't trust them."

Joe proposed sending Jack Downey and Dick Fecteau.

"They're strong, they're reliable, and they've been cleared."

Once summoned, the two men accepted the assignment despite its attendant risks. Theirs was a true sense of mission.

The plane took off on schedule, never to return.

⤝

THIS WAS THE WORST PART OF JOE'S JOB: PUTTING SOME one else's neck on the line. He had helped hand-select Downey and Fecteau and now they were presumed dead. He had no idea what had happened to the plane. All that was clear was that two young, healthy men with their whole lives to look forward to were gone. It was Joe's initiation into just how cold the war could be.

Needless to say, he couldn't talk to me about it. And I'm sure it didn't help matters when, at a Christmas party later that year, I asked Joe where "those two great new recruits were." He didn't let on that they had been lost on a mission. He explained offhandedly that they had been transferred.

But looking back, the signs were there that something was

wrong. Joe was working longer hours than usual. Even when he was home he seemed distracted. Quiet at meals. And sometimes, late at night, he'd be gone from bed. I'd come down to the kitchen to find him mixing his favorite snack, *ochazuke*, a tea-and-rice combination mixed with seaweed. (I guess we all have our comfort food.)

❧

IT TURNS OUT THE WHOLE THING HAD BEEN A SETUP. THE Chinese Nationalist agent had been apprehended earlier by the Chinese Communists and then coerced into sending the urgent radio message for a pickup. Chinese militia were waiting at the rendezvous point when the plane arrived, and shot it down. Jack Downey and Dick Fecteau were the only survivors. Captured, they were chained and placed in solitary confinement.

Each of their cells measured roughly eight by ten feet. Their accommodations consisted of a single cot and a latrine. Their cells had no windows; only one light bulb, constantly lit, served as illumination. Downey was shackled to his bed and not allowed to get up to relieve himself. They were allowed little sleep. When they did fall asleep, the guards would awaken them with a cold steel prod. Both men were subjected to interrogation and occasional torture.

It's reputed that when Dick Fecteau was tortured by the Chinese and pressed for the identity of his CIA colleagues, he simply rattled off the names of his teammates on the Boston University football roster. Satisfied, his captors ceased their interrogation.

Jack Downey kept in shape by running ten miles a day in place. (He counted the steps.) He was allowed to read only two publications: *Sports Illustrated* and the *Yale Alumni News*. It was through *Sports Illustrated*'s bridge column that Jack learned that Communist party leader Deng Xiaoping had been purged as a result of the Cultural Revolution—turns out Deng was an avid bridge player.

It would be years before news of their survival filtered out.

❧

CHAPTER IX

ONE HAND CLAPPING

I ONLY SEEM TO REMEMBER TOKYO IN BLACK AND WHITE. A labyrinth of alleyways and garish neon, cigarette smoke and wet leaves in gutters, all synchronized to the plaintive wail of a sweet-potato vendor plying the dusky streets with his wooden cart. "Yaki-imo! Yaki-imo!"

In 1954, Joe was assigned to Tokyo. Between Kamakura and Tokyo, we had a brief stay at the base compound at Atsugi. It was a tidy, isolated community with a military atmosphere and austere surroundings—an oasis of Americana. Atsugi was a unique experience for us, the first time I truly felt a part of the Agency family. Our English-speaking friends were just a doorbell away. We ate together, drank together, our children played together. We could all relax a little. After all, we shared the same secret.

Tokyo was another story. It was the big time, where the action was—the industrialists, government officials, labor unions, intellectuals, student groups. No more grappling hooks, sea captains and plane drops. And it was right about this time that the Liberal Democratic Party (LDP) came into existence.

Strongly anticommunist, the LDP was the critical component

of the Agency's efforts to counter the increasingly powerful Japanese Communist and Socialist Parties. These parties had forged a strong opposition to the mutual security pact between the U.S. and Japan and to the buildup of Japan's Self-Defense Forces, a key element to the balance of military power in the region. Thanks, in part, to the CIA's deep pockets, the LDP would go on to govern Japan for thirty-eight years.

In Tokyo, Joe headed up political action and propaganda for the station—fitting for Joe, as his mentor at SAIS had been Paul Linebarger, the guru of psychological warfare. His book, *Psychological Warfare*, published in 1948, is still considered the bible on the subject. He and Joe had stayed in touch after Joe's graduation, and Paul Linebarger and his wife were our houseguests in Tokyo. (In fact, I believe that Paul played a role in recruiting Joe to the Agency.)

Dr. Linebarger was a fascinating man and a prolific writer. In addition to academic works (which he published under his real name), he wrote a string of books that dealt with, of all things, Christianity and science fiction, under the memorable pseudonym Cordwainer Smith. He also wrote spy thrillers.

⤜

I COULD HAVE USED A FEW OF DR. LINEBARGER'S BOOKS myself. My days—no, months—in Tokyo were spent in bed, pregnant again, this time replaying Dvořák's Symphony "From the New World." Unlike in Kamakura, I had sense enough to position the bed so that it looked out onto the sculpted pine trees and bamboo thatches of our garden.

The bamboo hid an entrance to a bomb shelter and air shaft that John and David stumbled upon one day while playing in the garden. It seems that the shaft led to a dirt tunnel that ended up in one of the kids' bedroom closets, the other hidden entry to the shelter. The kids loved the idea of this secret passage into our house. I wasn't quite so crazy about it.

It could be that the time alone was just starting to get to me.

Like a character in a Hitchcock movie, I was alert to every prompt stillness. (Maybe the psychiatrist at Hopkins was right. My imagination was a little too vivid.) I'd flash back to the dill pickles and tuna fish sandwiches at Lexington Market. Or I'd stroll again along the Ala Wai Canal to Waikiki, sun so bright I'd have to shield my eyes. Did I miss a spot? That glorious frangipani bush next to our house on McCully Street? I'd go back and retrace my steps. I had time.

Joe would call, usually in the afternoons. It was always brief. Sometimes, he just wanted to hear my voice. His voice was a tonic—intimate, rough (but not gruff), unhurried, slightly Bostonian. (He called me "dea-ah.") I can almost still hear it. Almost.

Then the kids, home in a flurry from school, would scurry up the stairs to say "hi," looking as if they were straight out of a British boarding school. David would be in his gray flannel short pants, navy blazer, knee socks, and maroon tie; Mary, in her navy plaid jumper. Nothing beats a well-dressed child. Then they would run out to join their friends in our yard to play "Otete Tsunaide," Japan's version of "Ring Around the Rosie." Joe loved it—seeing his own children, boisterous in fluent Japanese, squealing and laughing with the neighborhood kids as they'd "all fall down."

❧

BY AGE ONE, JOHN ENJOYED THE RUN OF OUR GARDEN. "SCAVenging" would be a better word. He ate everything in sight: strawberries—any kind of berries—and flowers. But marigolds were by far his favorite snack. We knew we had to end his grazing, so we stuck him in a playpen in the middle of the garden—for hours. He loved it. Everything fascinated him—birds, bumblebees. (He probably would've eaten them, too, if he could have gotten his fat fists on them.) In fact, he spent so much time out there that he became a sort of legend.

One very snowy Sunday, one of Joe's office colleagues dropped by. Without even glancing, he queried Joe:

"Where's John? Outside in the playpen?"

❧

I'D TAKE BREAKS FROM BED, WALKING AROUND THE HOUSE TO keep limber, stopping to rest in different rooms. There was a lot to explore. Our home and garden covered an entire city block in the Meguro section of Tokyo. (I went back in 1985 and tried to find it. It's gone.) The previous occupant had been General Charles Willoughby, Douglas MacArthur's chief of Army intelligence.

The house seemed an uncanny reflection of our marriage: half Asian, half Western, all under one roof; well maintained on the outside, less so inside; in places austere, private, and almost forbidding, in others a riot of unexpected color; in all respects unique.

Joe would arrive home, usually late. I'd wait, cautiously, to read his mood. It wasn't easy. Being married to a spy is a little like being married to an actor or an adept poker player. Joe worked hard at not betraying anything. His skills couldn't help but follow him home—so I'd listen for that whistle or lack of it. I don't think he realized what a dead giveaway it was.

It was frustrating for both of us. He wouldn't (and couldn't) unwind, unburden himself, talk about an office conflict, get my second opinion. He kept me in the dark. I didn't even know Joe's title in Tokyo or where his office was located.

The nature of his work meant that all the frustrations and anxieties over operations stayed pent-up—until he would take them out on me. With Joe, it was always something. The fish should have been steamed, not grilled. The thank-you note should have been sent sooner. But why had I spent so much on stationery? From there it was on to our two favorite subjects: his mother and money. We would have saved ourselves a lot of trouble if we'd tape-recorded our fights so we could just replay them and go on about our business.

Joe kept telling me "not to fight fire with fire," but it wasn't fair—he always got to be the "fire." So I'd let him have it right back, recalling something my Uncle Clark had once said: "Bina, why be difficult when, with a little effort, you could be impossible?"

Impossible I sometimes was, but somehow we always managed a fragile truce. Joe would be there at the end of the evening, giving me a bed bath, body massage, and some kind of shot prescribed by the doctor. (He practiced on a grapefruit.)

I never knew what to expect from him. Once, he'd been away on a quick trip to Hong Kong and wasn't scheduled to return until the following day. I was standing out in our garden, when I felt his arm encircle my waist. I turned and smiled at him, thrilled at his touch. He smiled and kissed me. With that, he turned, approached his car, nodded to the driver who opened the rear door, and they drove off.

You see, when he wasn't terrible, he was wonderful.

⋘

A FEW MONTHS INTO THE PREGNANCY, I WAS GETTING FED UP with being in bed. A chance visit by a friend convinced me to get up after all those months and let nature take its course. (She was playing right into my own wishes.) I was up, already fully dressed, when the maid ushered in another friend. My friend gave me one hard look and said:

"Bina, what do you think you're doing? You get right back into that bed."

On that little happenstance hung Ann's fate.

Ann Sabina Kiyonaga was born on August 16, 1955, a lovely Sunday. She was simply beautiful. Tiny, fair with dark features— a female version of Joe. To this day she remains the best-looking of the Kiyonagas.

With Mary, David, John, and Ann, we now numbered six. It was worth the wait, and sacrifice, to have a good-sized family. Joe was a devoted father although sometimes his devotion veered on the manic.

We were all at a pool in Tokyo. Mary and David were splashing around. I was at the shallow end teaching Ann the dog paddle. John was reluctant to get in since he didn't know how to swim. Joe figured it was time that he learned. After coaxing John unsuccessfully to jump into his arms, Joe lost patience, pulled himself out of the pool and, unceremoniously, pitched John in. The only outcry was mine. I was furious. I thought it was a rotten thing to do. John weathered the incident fine, but I couldn't help but notice some passing resentment toward his father in John's eyes.

We were a handful and a houseful, especially when you added our house staff, "carefully screened" by the Agency. Because of security concerns, we didn't want help that spoke English. Joe double-checked the help by going up behind them and saying, "Fire." No reaction—they were in.

We'd left Tomi-chan, Ken-san and the ever spooky Mori-san behind in Kamakura. In Tokyo, we had Miki, the martinet; Nobu, the militant; and Keiko, our sweet, retiring cook. Miki, especially, stands out in my mind. She was fun—unusual for a Japanese housekeeper—and the children liked her. I marveled at how fast she could get the kids changed for bed. I later discovered her secret.

She'd line up David and John in their bedroom, and, fingers pointed in a mock bayonet, shout "Banzai!" With that, the boys would hop to attention and raise their hands over their heads as if surrendering, while Miki deftly removed their shirts and slipped on their pajamas. Ugly historical allusions aside, it was actually kind of charming.

You can't help but become a little attached to some of your servants overseas. A random act of kindness can do the trick. When John, our two-year old, wasn't busy foraging for delectable edibles in the garden, he would spend time visiting with Keiko-chan. (John's a great visitor.) I think he liked her better than he did me. She must have felt the same way about him.

On one of her days off, she asked me if she could take John for an outing. Surprised, I agreed, and watched as she meticulously dressed him in his Sunday outfit.

John was back about three hours later. When I queried him about his outing, he smiled, babbled something unintelligible, and I was reassured.

A week later Keiko-chan presented me with a gift—a tender

photo of her, still in her maid's uniform, and John. She had spent her time and hard-earned money to go to a portraitist to record her love for our son. Believe me, there's no better way to ensure a soft spot in a parent's heart.

Less charming was Nobu, the "very special" houseboy (the Agency's words) that they sent to us upon our arrival in Tokyo. He came highly recommended, and was good-looking and smart. Too smart.

Nobu had been with us about four months. The children liked him. I found him to be efficient and both of our maids had fallen in love with him. He was too good to be true.

One evening, I was upstairs reading to the children while Joe was meeting with a business contact in his study, downstairs. He thought he heard someone outside his study door; putting his finger to his lips, he quietly made his way toward the sound. When Joe opened the door, he was startled to find Nobu on the other side—listening.

More thorough checking of Nobu's bona fides revealed that he was a former Communist Party member, possibly a plant. You don't take chances; Nobu was sent on his way that day. Furious with his dismissal, he laid waste to our beautiful garden, scattering trash throughout, even stringing toilet paper from the trees (a sinister tactic he'd learned, perhaps, from disgruntled American trick-or-treaters). The place was a wreck. I guess he wanted Joe to "lose face"—who knows? perhaps even commit hari-kiri. But he didn't know Joe, who simply asked the maids to clean things up while he went inside to tend to more important business: the creation of the evening's martini.

❧

SECURITY IN THE HOME WAS ALWAYS OF PARAMOUNT CON-cern. I learned in Japan how something as ordinary as a telephone pad could cause problems.

A security man from headquarters would occasionally visit the station. He was not a technician; he was an examiner, adviser and lecturer. He would case the office, take note of any security shortcomings and recommend remedies. He would also speak to the wives, as a group.

I recall hosting a talk by a security man when we were living in Tokyo. About twenty wives were assembled in our living room. The visitor spoke to us for about forty-five minutes and asked if he could use the phone before he fielded questions from our group. I led him to the hall phone and left him to join my guests. He found what he was looking for (I doubt if he made a phone call) and continued his lecture, directed at me. He produced my telephone pad and cited twelve of my security lapses: the first twelve entries on my list. It started with Joe's office number and went down through the hierarchy of office wives, giving their names and home numbers. The security man explained how serious a breach mine was, how any guest, servant, or child could gain access to the list. He admonished me, and I learned my lesson well. I never put last names on subsequent lists. Instead I used an initial and scattered the names. Sometimes it was hard for me to figure out who was who.

✣

AFTER ALL THAT BED REST WITH ANN, I WAS READY TO FLING myself on Tokyo. If Kamakura had been ancient sword-making shops and snow falling on the ocean, Tokyo was camera billboards, gray concrete buildings (low for fear of earthquakes), jazz bars, drinking dens, pachinko parlors (the Japanese version of pinball), and tucked-away restaurants specializing in snapping turtle, whale sushi, and even raw horsemeat (called *sakura* for its resemblance to cherry blossoms).

The major intersections provided their own show. I speak of the Tokyo traffic police. They would stand on a small platform in the center of the road, decked out in smart blue jackets, peaked caps, and white gloves. No whistle was necessary. You took your cue from the gloves. They beckoned, waved, pointed, and pirouetted, like conductors, deftly orchestrating the flow of traffic.

One of my favorite areas was the Ginza, the Fifth Avenue of Tokyo. It housed department stores like Takashimaya (which, in fact, now has a branch on Fifth Avenue). You'd enter the store to the synchronized "irasshaimase" (welcome) of kimono-clad store clerks bowing in concert, and then thread your way through the Mikimoto pearls, handbags of knitted cypress bark, art deco bamboo table lamps and silver whiskey sets.

The Japanese are amazingly shrewd people. On the one hand they were the smiling, acquiescent vanquished; on the other, they were busily ripping off their unsuspecting conquerors. Shopping was a real experience in Tokyo—especially with Joe. His looks and height were deceptive to the unwary shopkeepers. Joe never betrayed his knowledge of Japanese. Inevitably, the salesgirls would size us up and charge astronomical prices. The situation would have been funny had we not been the victims. Joe usually intervened in impeccable Japanese, offered, and got, a reasonable price.

✄

JOE'S SKILL IN JAPANESE MAY HAVE COME IN HANDY WHEN WE were shopping, but it proved paramount when it came to recruiting agents. What's more, Joe had a certain way about him, a quiet reserve that I think the Japanese respected. He could sense where to draw lines, when to strike, when not to push. Joe understood more than just the Japanese language; he understood their body language.

Joe needed these advantages. His Agency colleagues in Japan had it all: contacts, credentials and charm—you name it. Many also had State Department covers, which were a huge boon. It sure opened more doors than being a "Department of Army civilian." Believe me, that didn't cut any ice with the Japanese. Joe might just as well have palmed himself off as a Fuller Brush man, it was that lacking in cachet.

The Japanese are big on titles—and education. Instead of, say, being in the political section of the Embassy, Joe wasn't free to say what he did or where. When questioned about his schooling, the University of Hawaii didn't resound with anything like the prestige of a Yale or Harvard.

Joe recruited many agents during his time in Tokyo—always concentrating on quality rather than quantity. (Of course, quantity didn't hurt.) He went from a GS-9 to a GS-14—roughly the equivalent of rising in rank from captain to colonel.

Chief among Joe's recruits was one of Japan's leading business figures of the day. Even though he's dead, I can't mention his name because his company is still well known worldwide. Joe first met this man in Washington. After arriving in Japan, Joe wisely renewed the contact. Although purely social at first, the big break came when the man announced his intention to travel to Latin America. Luckily, my parents had returned to Bogotá, Colombia, where my father was heading up the Point Four Mission (now known as AID). Joe cabled Daddy, asking that he roll out the red carpet for our friend and his traveling companions. They numbered six in all. Daddy met them at the airport, acted as tour guide, threw a party in their honor and even placed his cars and drivers at their disposal. (That Dale Carnegie course was still paying off.) This proved to be the turning point in Joe's relationship with the businessman.

Soon after his return to Japan, the same businessman headed a trade mission to China. Given his prominence and economic clout in Japan, the man had unique access to the upper echelons of the Chinese government and, through them, to crucial intelligence on the political and economic situation there. It was the perfect way for Joe to gather intelligence—through a third party already going to China on other business. The man amazingly agreed to relay information to Joe and refused to accept any remuneration for his efforts. He did it for the mutual benefit of our respective countries. The operation in China through this contact was so successful that it continued for years.

Joe's other focus was on thwarting the increasingly powerful Japanese Communists—that's where the political action and propaganda part came in. Joe had learned from Dr. Linebarger that an effective propaganda tactic was to find a trusted, respected and influential third-party "voice" (maybe a newspaper editor) to broadcast your (anticommunist) message or spread unflattering information to neutralize your foe. The key was that no one should ever be able to determine that *you* had provided the information.

Sometimes you just luck out when it comes to recruiting agents. Joe had been targeting a top journalist with the *Asahi Shimbun* for some time but had been unable to set up a meeting. He was reading the newspaper one day and learned that the man had taken ill. He was going to be in the hospital for a protracted period. With a few phone calls, Joe figured out which hospital.

Joe wrote him a letter, saying that he admired the man's journalistic prowess and would like to come by to pay his respects. Joe sent the letter but timed his visit to precede the note's arrival. Caught off-guard the man was simply too polite to dismiss Joe, and a pleasant conversation ensued. (I'm sure the letter's arrival later that day lent credence to Joe's sincerity.)

After that first meeting, Joe became a virtual candy striper. He'd visit the man regularly bearing magazines, delicacies, and best of all, prized articles from the PX—Camels, Dewar's and even powdered soup—nothing ostentatious, just enough to bait the hook. As a final push, Joe offered financial help to the man's family. (The hospital bills were piling up, with no income coming in.) As soon as the man recovered his health, a steady stream of articles—political, economic, and anticommunist—began appearing in the *Asahi*, the *New York Times* of Japan. Each morning, the throngs of Japanese commuters would read the thoughtful editorials on their way into work. The message was getting out, all right.

Of course, what better way to fight the Communists than to recruit one to your side? Joe targeted two high-level Japanese Communist Party members. These men were very special. They had trained under Mao Tse-tung in the Yunnan hills of Northern China and had returned to Japan—one to become the national leader of the party and the other his deputy. Word had it that they were becoming disaffected. Bull's-eye.

Joe cultivated them clandestinely, meeting in "safe houses"—locations maintained by the Agency for secret contacts or meetings. Food and drink were brought in, with emphasis on the latter. (I'm assured that no geishas were involved.) Joe could speak to them in Japanese, without an interpreter—a crucial factor in gaining their trust. I think, though, that Joe sometimes tended to out-Japanese the Japanese, especially when it came to being discreet and not jumping the gun. After many, many meetings, finally Joe asked them the question—would they collaborate? Chagrined, he overheard one of the duo turn to the other and mutter, *"Kore wa shibaraku deshita"!* or (loosely translated), "Good God, I thought he'd never ask!"

❧

SOMETIMES I, TOO, WOULD, UNWITTINGLY, OUT-JAPANESE THE Japanese.

One afternoon, I was browsing through Takashimaya, the department store. It had been a long day, and I, unfortunately, had worn new shoes. I just couldn't take it anymore. I peeled off my shoes.

As I padded off an escalator, shoes in hand, a saleswoman approached me. Glancing down at my stocking-feet, she looked at me quizzically, smiled and politely advised:

"It is not necessary to remove your shoes when entering a Japanese department store."

❧

JOE PUT ANOTHER SKILL TO USE IN MAKING CONTACTS WITH key Japanese figures of the day—his golf swing. The Japanese love golf. Thanks to a lot of downtime at the University of Hawaii, Joe had developed a good game. One of his first moves upon arriving in Tokyo was to join one of the leading golf clubs. It was while teeing off one day that he met the Chief Justice of the

Japanese Supreme Court. Neither collaborator nor agent but a good contact, and more especially a trusted friend, this gentleman helped fill Joe in on things economic and political. They were often joined in their discussions, over a glass of Old Parr, by the jurist's friend, the former president of Keio University. Both men were also tutors to the Crown Prince and really had their fingers on the pulse of the nation.

Then there was a cabinet minister who proved invaluable in providing Joe with economic intelligence. He really came to be fond of Joe; he even gave us a gift of a turkey for Thanksgiving. I was confused that Thanksgiving when the pungent aroma of fish began emanating from the kitchen. I figured it must have been the remnant of some previous dinner—that is, until we took our first bite. The minister had failed to mention that the turkey, like everyone else in Japan, was fish-fed.

⁌

THE MORNING AFTER THANKSGIVING 1954, JOE PLACED A copy of the *Strait Times* (Singapore) in front of me. He didn't need to explain further: the front-page story was about two Americans who had been tried and convicted of espionage by the Chinese. Their names: Jack Downey and Dick Fecteau. Jack had been sentenced to life imprisonment; Dick, to twenty years.

What? I thought they had been "transferred." I think Joe was relieved to finally let me in on what had been troubling him. Between drags on his Pall Mall, he explained how he'd suggested that Jack and Dick be sent as last-minute replacements on the mission out of Atsugi, how the plane had been shot down and the two men presumed dead. Joe and his colleagues had only recently learned that Jack and Dick were even alive.

Just two years earlier, I had danced with Jack Downey at that Halloween party. And now he was serving a life sentence in a Chinese prison.

Joe was relieved that the men were alive, although suffering

God knows what indignities in some Chinese prison cell. He doubted he would ever hear from them again.

≫

JOE WAS A BUSY OPERATIVE ALL RIGHT, BUT HE WAS JUST ONE of many. I don't know what his colleagues were up to, but I do know that they were pretty impressive. When I think back on Agency friends, it's these people who served in Japan that I think of first. Many are still my friends, after forty years.

One young colleague of Joe's was "Bronson," a Nisei from the West Coast. His father had been a renowned art dealer. Despite that, the whole family was interned after Pearl Harbor. Bronson gloried in things Japanese—he practically lapped up Japan with a spoon. He introduced our family to Japanese theater, Kabuki and Noh (the latter an excruciatingly slow and mannered style of performance that is downright weird). He also took us to Odawara, a town chock-a-block with lovely earthenware. I reveled in these casseroles. (I recently broke a large one and had to pay $175 to have it repaired. My recollection is that I originally paid about $8.00 for it). Bronson was so close to our family, that our children still refer to him as "Uncle Bronson."

And then there was "Floyd." A former Marine, he took Tokyo by storm—or at least his Great Dane did. Floyd would tool around town in his chauffeur-driven Toyo-pet (I'd like to say it was red, but I can't honestly remember) with the dog seated up front with the driver. Since both Floyd and the driver were smallish, the casual passerby might have though the Great Dane was doing the driving.

Postwar Japan attracted more than just an eclectic mix of Agency operatives. With the aftermath of the Korean War, the Chinese Communist threat and the Japanese Communist presence, Japan in the 1950s became the destination of an international potpourri of journalists, academics and entrepreneurs. All types of businesses were rushing to Japan to help in the rebuilding. My favorites were the newsmagazine and wire-service types.

Their Press Club on Shimbun or Newspaper Alley was the social center of Tokyo. Joe and I were there when Elizabeth Taylor and her then husband, Mike Todd, visited. She was so tiny that I couldn't catch a glimpse of her because of the throngs of reporters. Another time Joe and I were invited to a party at which William Faulkner was to be the honored speaker. Problem was he drank himself across the Pacific and was unable to stand, let alone speak.

More memorable than any celebrity was Father Joseph Roggendorf. He wasn't famous. (Although, years later when he visited the New York Stock Exchange, he was greeted in style by the ticker-taped message WELCOME FATHER ROGGENDORF) Nor was he a contact. He was a friend, a good friend. Dean of graduate studies at Sophia University in Tokyo, Father was a German-born, Cambridge-educated Jesuit. Articulate, debonair and knowledgeable, he provided Joe with wise and seasoned counsel on things Japanese. For a priest, Father was surprisingly urbane. (Then again, this may be the rule rather than the exception—I haven't known many European priests.) He even introduced us to sumo, that most tradition-bound, ritualized and revered spectacle of two humongous men, their hair in a knot, in silken thongs trying to tackle, smack, pummel, and shove each other out of the ring. It was great. Actually, I most enjoyed the *obento* lunch that we'd be served at the matches, a presentation of sushi and other delicacies in a lacquer box accompanied by a cold Kirin beer.

And, most important, Father was responsible for reconverting Joe to his faith. Although Joe'd been raised Catholic, he'd been baptized, made his First Communion, and that was about it

❧

JOE HAD SEEN TOO MUCH IN HIS CHILDHOOD AND DURING THE war—too much senseless suffering—to blindly make the "leap of faith" that I urged. He had questions. His would have to be a conversion of the mind and spirit. Father Roggendorf was the man to do it.

On Friday evenings, after a long week, Father would join our family for dinner. We always made it a point to serve American food; I figured Father'd had his fill of raw fish. We'd polish off the commissary-bought prime rib and baked potatoes with plenty of sour cream. Then Joe and Father would retire to the study.

Ours was quite a study, with walls paneled in lustrous dark wood, and deep leather-pillowed window seats under each leaded window. Even the tatami floor didn't detract; it just added to the unhurried, unaffected effect. It was a real man's room. And given that our predecessor had been Gen. MacArthur's chief of Army intelligence, I can imagine that that study had been the scene of many a late-night cloak-and-dagger strategy session.

I would go upstairs with the kids and leave Joe and Father to the privacy of the study. They would settle themselves into the two Danish-inspired, low-slung, teak chairs. The sturdy chairs were a perfect complement to their robust occupants. Over coffee, cognac, and cigarettes, Joe and Father would converse. At first, their conversations centered on the secular—Japanese politics and Winston Churchill's history of World War II. They eventually moved on to Teilhard de Chardin, Hilaire Belloc, and G.K. Chesterton—theologians who grappled with the essential mysteries and paradoxes of faith. Late into the night, Joe would question Father about the concept of the Trinity; the Mystical Body; and, the miracle of the Eucharist.

Our marriage was the true benefactor of these late-night talks. Even though Joe and I were both Catholics, there had always been a gulf between us on the issue of faith. Joe had seen my faith as almost superstitious. Now he understood there was far more to it. We would always approach the world differently, but now, at least, from the same perspective.

⤜

TOKYO AND I WERE GETTING ALONG JUST FINE. JOE AND I were, too, but I couldn't help having the sense that while Joe was trying to contain Communism, I was . . . shopping? Not exactly the kind of partnership I'd envisioned. It's not that I wanted to work for the Agency, too (one spy per family is plenty), but I thought maybe I could do something to help Joe. Enough sightseeing by myself—I decided to try to learn something about the culture and mix more with the Japanese. Maybe, in the process, I could initiate some contacts for Joe.

Thus began the classes in ikebana, the Japanese art of flower arrangement, especially the *sogetsu* style that used everything from wheat stalks to prickly pears to glass tubes. The classes were instructive, and fun, since several other Agency wives joined in, as well.

Then on to *sumi-e*, the art of Japanese calligraphy, done in charcoal ink on rice paper. I was privileged to be tutored by Oi-san, a Japanese National Living Treasure (artists subsidized by the government because of their renown). The key element in *sumi-e* was to dip the paintbrush in the thick black ink and write the character all in one stroke. The result was an energetic but controlled melange of jet black, which eventually dissipated into gray streaks.

Pottery classes completed the picture, taken from none other than Hamada, another National Living Treasure, who produced earthen-tone bowls of flawless imperfection.

All the while, there were the inescapable tea ceremonies, with their glacial traditional rituals, i.e., turn the cup, then drink the tea in three sips. Not exactly my cup of tea.

❧

THE JAPANESE CAN ASTOUND YOU WITH THEIR THOUGHT-fulness. My sumi-e instructor, Oi-san, hearing of Joe's fondness for "Otete Tsunaide," presented him one Christmas with the words in calligraphy, the stark black letters on an off-white background, mounted on rough beige paper, remounted on rougher taupe paper—and all framed with simple brown wood. It hangs now in my son John's law office.

Hamada invited the children to his home to teach them the

art of ceramics. (Picture I.M. Pei teaching your four-year-old how to design a dollhouse.) He later sent them their clay bowls with a note:

"Here are your clay bowls. Hold them in your hands."

⁓

MY EFFORTS STARTED TO PAY OFF. IN REAL ESTATE, IT'S location, location, location. In the intelligence world, it's access, access, access.

Through a friend from my ikebana class, I was invited to join the Women's Round Table Club of Japan, a small social club for the wives of Japanese and American government officials and captains of industry. Mrs. Yamashita, whose husband headed up Japan Airlines, was a charter member. Her unerring taste in kimonos extended to her choice of gifts. Hers to me—an old Imari blue and white vase, softly splashed with a suggestion of bamboo—is still proudly displayed in my living room.

I was also asked to be cochairwoman for the annual dinner/dance at Saint Mary's International School, that David attended. The dinner/dance was a highlight of the season, a perfect chance to mingle. I decided to hold some of the dance committee meetings at our home to put things on a more familiar setting. From there, it was just a step to dinner party. That's where I could be of real help to Joe—by setting the stage for the true drama of his mission, agent recruitment.

These dinners were an important extension of Joe's working day—a chance, in a relaxed, comfortable setting, to cultivate promising contacts. The experience had to be seamlessly discreet, private, and in a manner befitting the level and sophistication of the target contact. (Fighting Communism was a tough undertaking—someone had to plan the menu.) I looked to a local institution, Chinsanzo, to learn the props.

Chinsanzo was a restaurant on the outskirts of Tokyo. Rather than your own table, you had your own cottage. You'd be led, over a watered stone path, by women in kimonos, each guest equipped with a unique form of illumination: a small cage of fireflies.

The cottages were sprinkled over the hillside. Small, one-room versions of the typical Japanese home, each cottage boasted its own seasonal scroll, complete with flowers in the tokonoma (a special recessed area).

I could do that.

Joe and I would receive our guests in the living room for cocktails. The living room looked out on a formal Japanese garden, through French doors. At dinnertime, kimono-clad maids would summon and lead us through the garden. Each guest was supplied with a bamboo cage filled with the requisite fireflies to light his way. Additional lighting was provided by candlelit Japanese stone lanterns dotting the garden.

Upon arriving at the Japanese wing, the sliding doors would be opened to reveal a traditionally set floor-height table, surrounded by navy-and-taupe-striped rough cotton cushions on the tatami floor. The guests were asked to remove their shoes, and sandals were provided. Situated at the far end of the room was the tokonoma where a seasonal scroll and flower arrangement were displayed. The motif might be cherry blossoms in the spring or quince in the fall. My favorite was pine branches and red roses during the winter.

Everything, in turn, set the stage for the meal. Much of Japanese cooking centers on the visual and the method in which things are served. Our Japanese chef would be seated directly behind the table where he cooked and directed the serving of various courses. Our table seated as many as ten guests. Any more would have detracted from the beauty of the service.

The usual Kiyonaga fare included sashimi; hot and sour soup (replete with shredded eel); tempura (batter-fried shrimp, leeks, and mushrooms); Korean fire meat ("Genghis Khan"); a broiled filet marinated in a teriyaki-ginger sauce and coated with sesame seeds; white rice; *gaikao*, or pickles—all followed by bean-curd ice cream and coffee.

We usually included one other Agency couple. "Nick" and

"Theresa" were a great addition. Nick was a Harvard-trained law-yer who was serving with NATO in Paris when he met Theresa in Paris. She was a press illustrator for the fashion houses when they married. On their honeymoon on the SS *United States* they could happily clink their champagne glasses without a care in the world; no fear of a second *Titanic*. Safely stowed aboard was Nick's gift to his bride—a brand-new ketch that he'd had custom-crafted in Norway.

I was never privy to what sort of recruitment plan Joe had in mind when he'd suggest that we entertain. He would handle the recruitment; I would handle the dinner—and set the scene for a welcoming, special time in our home.

Sometimes, often weeks later, Joe would indicate whether or not the dinner had been a success. He never let on which guest he'd recruited.

As our guests would leave for the evening, you could hear the "click-clock" of the night watchman as he went on his rounds, banging two wooden blocks together to ward off would-be cracker thieves.

❧

THESE DINNERS WERE IMPORTANT, BUT I WOULDN'T HAVE exactly called them fun. They were just too purposeful with their hidden agenda.

When Joe and I wanted to be alone, really alone, we'd head off to our favorite spot in Tokyo: the bar at the Imperial Hotel. No contacts, no careful conversations, no talk of the kids. Just us. Designed by Frank Lloyd Wright, the highlight of the hotel was the spare, rustic but inviting bar. (The old Imperial has since been torn down; the new Imperial is an impressive but unimaginative skyscraper. The only part of the original structure that remains is the bar.)

We're always complaining that the Japanese take our ideas and improve upon them. It appeared to me, looking at that bar, that Frank Lloyd Wright had returned the compliment—he'd taken some ideas from Japanese design and done them one better.

Our dinner at the bar was usually as spare as the surroundings: raw oysters and stone-cold martinis. I recall one particular Friday when we cut our meal short to attend a special event at the small, private theater in the basement of the hotel: the Japanese premiere of *The Bridge on the River Kwai*. It was an invitation from a contact of Joe's, and a real treat. Living abroad, I missed having the occasional night at the movies. And the 1950s had produced some of the best American films, like *Twelve Angry Men* and Joe's personal favorite, *High Noon*.

So it was with anticipation—but some trepidation—that Joe and I went to see *The Bridge on the River Kwai* in 1957. We knew little of the movie except that it involved depictions of cruelty by the Japanese toward British prisoners during World War II.

We were entranced from the opening scene—the cheerful heroism of the prisoners whistling the *"Colonel Bogey"* march—but started to grow uncomfortable with the mounting silence in the theater.

As the movie went on (and on), I glanced around and confirmed my suspicions—I was about the only Caucasian present.

I really started to squirm when it became clear that the British hero, a prisoner of war, was going to make the supreme sacrifice. After months of cruel and tortuous subjugation by his Japanese captors in building the railroad bridge—critical to the Japanese for supplying ammunition—the hero throws himself on the explosive detonator at the last possible moment. He blows himself up, along with the bridge. The cruel Japanese had been defeated once again! (You could have heard a chopstick drop.)

And then, as the credits started to roll, the entire audience stood, as one, and broke into thunderous applause.

❧

MAYBE IT WAS SEEING THE JAPANESE REACTION TO *THE Bridge on the River Kwai* that prompted Joe to "go Hollywood." I found out recently from a colleague of Joe's that Joe had spearheaded an effort to produce a Japanese film, funded by the Agency, to counter some of the rampant Communist propaganda in Japan. It was called *I Was a Prisoner in Siberia*, and it depicted the torture inflicted by the Soviets on Japanese prisoners during the war. Apparently, it enjoyed some theatrical success and caused quite a stir. It was psychological warfare at its best. Dr. Linebarger must have been proud.

᪆

JOE WAS FINE IN THE SPY ARENA. BUT HE COULD NEVER SEEM to relax; he was always on. At work that was a given, but even in a quasi-social situation he was on his guard. It was as though he couldn't accept that he'd been accepted, as if he was afraid he'd be found out. In his own mind, he just couldn't seem to shake that "town bastard" image.

Not that anyone would have guessed. Joe at a party was worth watching. To see him leaning back, palming a drink, smiling quietly, you sensed an air of probity. He took his time whether it was responding to an inquiry or raising an eyebrow. Maybe that was his greatest coup. He had just about everybody fooled. Even me, sometimes.

One evening we were invited to a minister's residence for drinks. We arrived at his surprisingly unassuming home surrounded by the inevitable gated wall as well as numerous chauffeur-driven, black town cars. It was going to be a big night. Joe led the way across the stepping stone-path. I dawdled to admire the garden of well-trained evergreens surrounding a minuscule pond. I was about to call Joe's attention to it when I glanced at him. He'd already reached the door, his back to me. Suddenly, almost imperceptibly, Joe took a deep breath (more like a groan) and hitched his shoulders.

I never let on that I'd seen him, but it made me sad.

᪆

JOE COULD BE EXPLOSIVE, BROODING, OBSESSIVE. AND CRIT-
ical. But he was never small.

In 1956, we left Japan for a two-month leave back home. A
superior of Joe's gave a large farewell party for us. They offered
to do the same when we returned. I declined—bad move. Joe
was never one for office politics, but I should have known better.

It seems I'd offended them, and they in turn, chose to offend
Joe. On his fortieth birthday, he was passed over for a promotion.
The worst part of it was that he had to face a "surprise" party
that Halloween night. (How anyone thinks they can "surprise" a
spy with a party is beyond me.) As we dressed for the party that
night, Joe mentioned offhandedly that he'd been passed over.
I have to hand it to him. We went to the party and Joe congrat-
ulated several of his colleagues who had received promotions
that day.

❧

JOE WAS TURNED DOWN FOR A PROMOTION BUT GOT GOOD
news, indirectly, from another front. Additional word had filtered
out about Jack Downey.

He had endured two years of solitary confinement and mer-
ciless torture. Yet, Jack Downey still had the presence of mind
to seize an opportunity to get an important message out
through a U.S. Air Force pilot being released from the same
prison. The gist of his message—that he knew his captors might
overhear—was that the Chinese knew all about "Sam Murray."
The officer assumed that he was carrying news of a friend, when
in actuality Jack was referring to the biggest operation against
the Japanese Communist Party. Its code name—Operation
Samurai—had been coined by my husband. When Joe and his
colleagues got the message, they discontinued the operation on
the assumption that the Japanese Communists were privy to
the same information.

Jack Downey was still on the job.

I knew none of this until twenty years later, when Joe told me from his bed at Sloan-Kettering.

❧

WE WERE GOING ON OUR FIFTH YEAR IN JAPAN WHEN WE GOT our second visit from Joe's mother. This time, she called to let us know she was bringing *her* own mother. This I had to see. Joe would be meeting his grandmother for the first time; the kids, their great-grandmother. It was a big event.

Our two guests arrived on a lovely spring evening at dusk. I glimpsed them through our stained-glass library window, the kids kneeling on the window seat next to me.

Joe helped his mother and grandmother out of the car. His mother was in a dark suit; her mother, in a severe kimono. I learned later that she was in mourning for her husband who had survived the atom bomb in Hiroshima, only to die a few days later from a heart attack. Thinking back, Joe's grandmother put me in mind of Mother Teresa. Same size, same build, same smile. She was ninety-two years old.

She approached me with a faltering gait. She looked so fragile that I hoped she'd last the weekend.

I went to greet her in my best ceremonial style—bowing as I approached. The old lady surprised me. She stuck out her hand and bade me hello—in English. (She'd lived in L.A. for years.)

As I welcomed her to our house, I inquired if there was anything I could get for her. Lighting a cigarette, she replied:

"Do you have any bourbon?"

I was starting to understand World War II.

❧

I THINK WE ALL COULD HAVE USED A DRINK. JAPAN CAN GET on your nerves. It's exquisite—but exhausting. I was feeling strangled by the silken constraints of Japanese ways, the incessant gentle politeness. I needed air. We'd been there almost six years.

Our children spoke English with a Japanese accent; we were becoming critical of things back home. But the deciding factor was when a visiting Agency official jokingly queried whether Joe was contemplating taking out Japanese citizenship.

Just before our departure, a group of Nisei friends gave us a dinner. Among themselves, they were convinced that if anyone could break out of the Far East mold and switch to another global area, Joe could. They dared him to give it a try. Joe had done well in Japan; his future with the Agency was assured in Asia or on the Japan desk at CIA headquarters. Notwithstanding, he decided to take the dare, with some encouragement from me.

We had one last hurrah, just the two of us. Joe and I returned to one of our favorite haunts—a hot spring ryokan, or traditional inn, in Hakone near Fuji. The straw-matted room was virtually bare so as not to detract from the vista of the mountains outside the windows.

Dinner was served in your room, and it was then on to the communal hot bath among the rocks, where the waitress would gently float our cocktails to us on a bamboo tray. The guests walked back to their rooms in their padded kimonos—the steam rising off bodies in cold weather—like some surreal pajama party.

How do you end a perfect evening? A massage. So the masseuses were summoned.

Joe had tutored me on the acceptable protocol: we were to await the masseuses in the buff. I was leery. Joe explained that the masseuses tended to be older, blind women. (To me it was a demeaning, but cruelly practical, use of their heightened sense of touch.)

There we were, both stretched out buck naked on our futons, when in walked our so-called masseuses—a young virile man and an attractive perky young woman. They sure didn't seem blind to me. The woman made a beeline for Joe and the man for me. Joe rose from his bed. He gave the man one look and that man was out of there. Joe could be pretty imposing—especially when he was stark-naked. For some reason, he wasn't as angry with the woman.

So it was with a laugh—and fondness—that, in 1958, we said good-bye to Japan.

CHAPTER X

❧

PLACE CALLED HOME

I WAS NOSTALGIC FOR THE ORDINARY.

After six years abroad, I couldn't wait to go grocery-shopping, take the kids ice-skating, crinkle open a Reese's Peanut Butter Cup, visit with *English*-speaking neighbors (relegating charades back to the living room). I was ready to go home—back to the land of dancing and cheese, neither of which you find much of in Japan.

It all sounded kind of—exotic.

What I really had missed overseas were certain things that I never even knew I liked. Mainly foods: coffee at Montgomery Donuts; a frozen custard at Bethany Beach; eggs and bacon at Bethesda's Tastee Diner; Hellmann's mayonnaise, cottage cheese and softshell crabs. (There are no substitutes for these last three anywhere in the world.) But my true yearning, my real ache, my "Rosebud" was for something even more prosaic: mashed potatoes and gravy.

❧

WHEN I WAS GROWING UP, DADDY, AS A SPECIAL TREAT, would take me to dinner at Pappa's, on the corner of Baltimore's

Poplar Grove and Edmondson Streets. Looking back, it qualifies as a "greasy spoon." But to me it was "four stars," with its red-checkered oilcloth table covers, Depression-era Coleman lanterns and industrial-size salt and pepper shakers. The master chef—Pappa himself—would occasionally join us. He'd plop down in our regular booth, the stains on his apron betraying the specials of the day. We didn't even order. When Pappa saw us coming, he always saw to it that two hot roast beef sandwiches with mashed potatoes and gravy were dished up and waiting.

✁

I REMEMBER ARRIVING AT WASHINGTON'S NATIONAL AIRPORT and being struck by how big and loud Americans seemed, just like their cars. And I loved it. I loved the way they unabashedly slouched—very *un*-Japanese. And how friendly and open they seemed. "Hey, how ya' doin'!" No more guarded conversations. No more garbled conversations. I jumped right in, engaging in an overzealous conversation with the customs inspector—by the time I was finished he knew each of our children by name, age and hobby. I couldn't help myself—my guard was down, way down. We were home.

But what was home? As far as the kids knew, it was Japan. And Joe's memories of "home" were very different from mine. His island "home" didn't have impromptu neighborhood games of "red light/green light" or sidewalks chalk lined with crooked hopscotch boards.

I don't think Joe's mind was on hopscotch or mashed potatoes as we landed in Washington. He was anxious, worried that he'd made a bad career move. A big one.

Joe had secured a future for himself with the Agency during his first nine years. If he'd chosen to continue to concentrate on Japan, he'd have returned to Washington as the deputy branch chief for Japan at the new CIA headquarters outside Washington in Langley, Virginia. (In Washington, if you work at "Langley," you work at the CIA.) But Joe had decided to gamble. He'd taken that dare at our farewell party and now was back to see what the Agency could offer in another part of the world.

Figuring that out would have to wait. First we had to start what would become a homecoming ritual: a visit to my folks' home in Arlington, Virginia, for pot roast with frozen peas and pearl onions, glazed carrots, scalloped potatoes, and buttered Pepperidge Farm rolls. Everything was cooked—actually, slightly overcooked—but tasted great to me. We were definitely Stateside.

Then the next day it hit me like a wet newspaper on concrete: we were back and on our own. No Agency driver to shepherd me on my chores. No icebox stocked with the latest from the commissary. No drive up the cobblestone path to check out our new quarters. No more East and West wing; no more sliding rice-paper doors. No wings, no door. In fact, no home!

The search was on, this time for a place to buy. But given our last search for a place in Washington, Joe and I weren't exactly optimistic. Fortunately, Americans tend to detest one enemy at a time—and in 1958, it was the Russians' turn.

Mrs. Murphy, our realtor and mother of an Agency wife in Japan, suggested our first stop: 3911 Leland Street in Chevy Chase, Maryland, a straight up-and-down colonial. Joe and I took a look around. He headed down to the basement, I to the attic. We met in the middle a few minutes later. We didn't have to say much—we knew. The house just felt comfortable, and, as Mrs. Murphy put it, "It's a good address." Plus, the price was right at $27,500.

The house had been built in 1925. That cinched it for me: we were the same age.

Joe and I made a "lowish" bid. By chance, we learned that another Agency couple had seen the house that same day and had also made a bid. They were an exceptionally attractive couple—and they had money.

We didn't stand a chance.

But more than anti-Japanese sentiment had changed; Joe had. He'd learned a lot about people since our last disastrous apartment search. Unbeknownst to me, while I was touring the house (and redecorating every inch of it), Joe was in the basement talking with the owner, a world-traveled *Time-Life* correspondent. They discovered they had mutual interests, and even a mutual friend—a journalist in Tokyo.

We got the house.

When we moved in, I remember thinking that the neighborhood houses looked like a row of stalwart, well-tailored matrons, staring out impassively with their shuttered window-eyes. I think our house was glad to see us: we were immediately embraced by its varnished paint smell, the knowing creaks in its floorboards, the roar of approval from its blazing fireplace. I give Joe credit. He's the one who dreamed up the idea of the broad black-and-white-striped tiled hallway. It took him back to Fred Astaire in *Top Hat*.

We'd been through six sets of living quarters. Finally, we had a place to come back to. A musty basement where we could track the kids' heights with pencil marks. A fresh palette. Our ground zero.

A home, even a sense of home, is important to the CIA family. Overseas, it's a whirlwind ride through first school days, raw octopus tentacles and befuddled store clerks. It's great, but you have no sense of permanency, no sense of community. In the end, it's not yours. No matter how many friends you make, how good a school you attend, how many clubs you join, it's just not the same. You can't help but feel like a welcome guest at a very gracious hotel.

Like good Japanese, we seemed to spend more time working in the garden when we moved in than on the house itself. First, we planted a stand of bamboo along the back fence and a reddish, somewhat droopy Japanese maple in the sunny part of the yard. (It missed its homeland.) Next to the small, flagstoned pond—which we stocked with lily pads and speckled *koi*, Japanese goldfish—we ceremoniously installed a vintage stone lantern we had bought at a Buddhist monastery in Hokkaido. It consisted of natural, asymmetrical stone slabs piled on top of one another—unique. (The whole effect was somewhat spoiled when one of the kids commented that it resembled a Mexican with a large sombrero taking a siesta.) And finally, there was our oyster-white pebbled driveway, a pedestrian allusion to that zen masterpiece, the twelfth-century rock garden at Kyoto's Ryoanji Temple. (We stopped short of watering and raking the driveway each morning, in the Buddhist rock-garden tradition.)

Along with our sense of home came a sense of community that had tracked us from Japan. Literally. Five of Joe's colleagues from Japan ended up buying homes within a two-block radius. Like former inhabitants of some strange island, we understood each other with a nod. And, more important, our kids were already friends.

Not that the kids didn't have their problems.

School was no fun for Mary. She would trudge home some days from Our Lady of Lourdes grammar school in Bethesda; her face said it all. Not even our cat, Peter, a Siamese who would wait for the kids to come home from school in our mail basket, could cheer her up. I'd watch Mary come up the driveway. If she seemed happy, I was fine. Otherwise, I'd mirror her mood and try to get her to talk about what was wrong. As it turned out, no one would sit with her at lunch. She'd entered school in the middle of the year, and the cliques had already been formed. One afternoon, she came home and broke into tears. Enough. Now we were both crying.

Without telling Mary, I made an appointment with her teacher, a nun, for the next afternoon. I explained the situation, but Sister was less than sympathetic. A little snippy, in fact. Finally, I asked if there was anything she occasionally did to make a student feel special?

"Well," she allowed, "erasing the blackboards after school."

I "suggested" that perhaps Mary might be selected for that honor. Sister didn't promise anything. But the next day, it was a smiling Mary who walked up our driveway. She told me she'd been chosen that day to erase the blackboards. And, sure enough, the next day, she had plenty of lunch companions.

I felt I owed that nun something, so when Mary came home

at Christmastime with big news and a strange request, we bent the rules. Mary was excited that she had been asked to provide a Japanese Christmas carol for an upcoming school pageant.

Now that's pretty tough, considering that Japan, a Buddhist country, doesn't celebrate Christmas. But Mary was so happy to be part of the pageant that I couldn't send her back empty-handed.

I can still picture Joe as I asked him to "just make one up—who's going to know the difference?" He was seated on our brown-tweed couch, enjoying his steaming mug of Wilkins coffee. He thought the idea was ridiculous, and I did, too. But Joe was a good sport about it. We selected "Joy to the World," which translated into "Ureshii-ku O-Christo ga Kita."

Festival night arrived and there was Mary on the stage in her kimono (the school's token Japanese) with her classmates happily belting out Joe's "traditional" Japanese Christmas carol.

❧

TWELVE YEARS LATER, WHEN WE RETURNED TO THE STATES after living in El Salvador, Paul, our youngest, entered Our Lady of Lourdes as a first grader. Joe and I dutifully attended the school's annual Christmas Festival. Imagine our chagrin—and embarrassment—when the nuns trotted out that famous Japanese Christmas carol, "Ureshii-ku," complete with words so everyone could join in. I remember Joe looking around, hoping no one from the Japanese Embassy had kids in the same school. The experience still haunts me as I now have grandchildren coming up who could possibly attend Our Lady of Lourdes. Do you suppose?

❧

IT WAS IMPORTANT TO ME TO DO IT UP RIGHT IN OUR NEW home. A kitchen that churned out the delightful scent of a roasted rosemary chicken, a living room that made you want to "throw another log on." I wanted Joe to know the fun of a newspaper delivered each morning, of having a cup of coffee before anyone else was up.

Soon after we moved in, we gave a bathroom-warming party. We'd converted a downstairs closet into a miniscule bathroom. Our friends were to help us decorate it. The color scheme I had chosen was gold and orange.

The day of the party, while I was out shopping, our friend "Cindy" slipped in the back door and painted a mural on the bathroom wall and ceiling. It was a tremendous tree festooned with gold fruit (luminous cutouts that she pasted on the wall.)

We had imaginative friends. Their gifts included a handmade gold ceramic mirror, an ivy-encrusted gold toilet plunger and a luxuriant orange bush that occupied the place of honor beneath my favorite Brazilian treasure—a brass sink fashioned in the shape of a shell. A starfish controlled the cold water, a sea horse the hot. Our friends' generosity proved to be almost too much. The bathroom was crowded but the party was a great success.

Soon after, Joe and I started a tradition. Our backyard boasted a really spectacular double cherry tree. The tree was the occasion for our spring party—the Cherry Blossom Viewing Festival. We had to time it just right and usually at the last minute. The only predictable quality of cherry blossoms is their unpredictability. Our menu was always steak tartare, asparagus with hollandaise, caviar and cream cheese on toast, pâté, Camembert and strawberries with whipped cream. Every drink sported a cherry blossom.

It made me happy to see Joe lingering over our new-found normalcy. It made me happy to see him, period. His regular hours were a relief. No more 3:00 A.M. phone calls or hurried departures from school plays. No more unexplained disappearances. I knew it wouldn't last; after an overseas tour, Agency employees normally spend two years in Washington, getting back in touch with headquarters. At some point we'd have to bid Leland Street goodbye.

But for a while I could relax. I was still discreet about what my husband did for a living; I was told to say that he worked for the Department of Defense. Security wasn't quite the same concern it had been overseas. I just didn't get the feeling that Mr. Turner, our repairman, could be a Soviet plant surreptitiously surveying our house.

It was more relaxing for us, but Joe was bored. He was assigned to a generic staff job while he roamed the halls looking for an overseas assignment. After all that excitement in Japan, all this "normalcy" was driving him crazy.

Every day, Joe came up with a new idea. He'd quit the Agency. He'd go into business—be a coffee broker (shades of my father's tales of Brazil). Or buy a franchise—some new thing called . . . McDonald's. If all else failed, he and our sons could always stage a bank heist to keep things going. At last, he'd be able to put his knowledge of disguises to use.

I thought back to Japan, when he had received a letter from someone representing the sugar interests in Hawaii suggesting that he return to the islands and run for the U.S. Congress. Joe turned them down. From where I stood, that seat in Congress looked pretty good.

Sitting it out at headquarters must have been especially frustrating for Joe because this was the era of some of the most daredevil Agency maneuvers. Only a few years earlier, the CIA had engineered one of its most famous (though some would say infamous) covert operations. It had been a real coup, literally. The Agency had ingeniously staged a fake "invasion" of Guatemala, complete with frantic radio broadcasts announcing the advance of the invading forces. (I seem to recall a similar tactic being put to work in some cowboy-and-Indian movies.) The left-leaning president of Guatemala, Jacobo Arbenz, scampered off to safety, allowing for a takeover by a more friendly (read: U.S.-leaning) government.

The Agency seemed to be everywhere, fighting its murky war on Taiwanese beaches, Congolese militia meetings, Iranian Parliament sessions and Berlin underground tunnels. It didn't bother me that Joe was doing battle with a desk at Langley. It meant he had more time to be at home—good news for the kids and me.

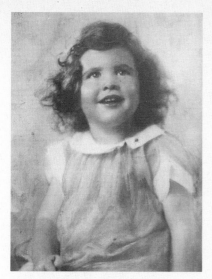

Bina at three years (1928).

Joe at three years (1920).

Kaunakakai, Molokai Grammar School, ca. 1923. Joe is in the back row wearing a white cap. He always did have style!

Bina with friends. Bogota, Colombia, 1944.

Mr. and Mrs. John Clayton Cady—Mother and Daddy. Bogota, 1944.

"M" COMPANY

"M" Company, 1943. In the back row again! Joe stood head and shoulders above the rest of his outfit.

Joe with a friend. Italy, 1944.

Bina was smitten! University of Michigan, 1946.

Mr. and Mrs. Joseph Yoshio Kiyonaga.
Washington, D.C., July 1947.

Our first CIA party was at the base in
Atsugi, Japan. Halloween, 1952.

Mother, Grandmother and two nieces in front of Daibutsu, my good friend the Buddha. Kamakura, 1953.

We were finally comfortable. Joe and me with our young family—David, John, Ann, and Mary. Tokyo, 1956.

With Helen Hayes. Brazil, 1962.

Joe and Paul. Bethany Beach, 1964.

To Joseph Y. Kiyonaga
with best wishes   Lyndon B. Johnson

Joe greeting Lyndon Johnson, El Salvador, 1968. Note the inscription from LBJ to Joe.

Entertaining at home. Marina Sánchez (the President's wife) to my left; Ellen Castro (the Ambassador's wife) to my right. Note Darcy Pentaedo's portrait of Ann above me. El Salvador, 1968.

The Kiyonagas: Joe, David, Mary, Ann, John, Bina and Paul. El Salvador, 1968.

Manuel Noriega and Joe. Panama, 1974.

The Kiyonagas at home. Panama, 1975.

He decided to dedicate some of his extra time to doing repairs around the house. Bad news.

Joe could orchestrate daring midnight skylifts deep into Red China, thrust and parry in the shadowy world of "psy-ops," discreetly press a foreign cabinet minister for sensitive military intelligence. But just try to get him to fix a toilet.

I think Joe's initial stab at handiwork and home repair started that first Christmas on Leland Street. Financially, things were tougher for us than they had been in Japan. There, the Agency had paid the rent (now we had a hefty mortgage), the salaried staff (now nonexistent), and footed part of the food bill (entertainment costs). It was different now. We missed those rents from the Hawaii houses more than ever. It would be a lean Christmas—we'd make, rather than buy, gifts for the kids.

I raided Bruce Variety for every fuchsia bow and ribbon and sparkly trinket they had, and then used the stuff to decorate an old yellow bulletin board I had found in the basement. Mary, now busy with school activities, loved it. I rustled up an equally glittery headband for Ann. Joe set about building a table-and-chair set for the boys. This was accompanied by much late-night sawing, an incessant—and usually futile—whirring of a tape measure, and innumerable trips to Hechinger's, the hardware store. He eventually quit after making a rather rickety chair, and bought the boys toy guns instead. (Who gives their sons chairs for Christmas anyway?) Joe was one lousy handyman.

He was, however, a determined handyman. Joe insisted on fixing the toilet to "save money," and even went out and bought *The Better Homes and Garden Guide to Home Improvements*. Huge, huge mistake. He not only bought the book, but bought the tools to go along with it. How sad.

The first to go was the childrens' toilet—in a veritable geyser. Eventually, a cadre of weary, world-wise plumbers, plasterers and painters got it right.

Then there was Joe, the painter. The house was white, inside and out, but I felt that at least our upstairs sitting room would be more attractive painted a different color. Pale blue—I emphasize the "pale"—was the color we chose. I suggested that we hire a painter.

"Why hire a painter?" Joe asked. Easter weekend was coming up. Joe would do the job himself on Good Friday—a sad enough day without Joe's dilemma.

Joe tackled his paint job from the wrong direction. He bought blue paint and worked backward from that, adding unbelievable quantities of white. Poor Joe. While I was ferrying the children back and forth to church, he was busy driving back and forth to the paint store for more white paint. Rather than face the situation head-on, I chose to drop by a neighbor's house. I noticed our green station wagon whizzing between our house and the paint store several times. Finally, Mary whispered in my ear:

"I wish Daddy would stop going to the paint store. It's making me nervous."

We ended up with a lovely pale blue sitting room and gallons of blue paint in the cellar. Word got around of Joe's painting prowess. I recall some Agency friends coming down from Boston to visit us one weekend. As I was giving them directions on the phone, the wife interrupted:

"Don't worry, Bina, we'll just look for the blue house."

The whole thing made me feel bad, especially since I had surprised Joe with a special outfit for his new home-repair career: matching blue denim workshirt and pants. The almost continuous washing (that paint again) rendered the whole outfit a dull, blea-chy white.

The new outfit, in turn, almost led to a career change. Joe was picking something up at High's, the convenience store (probably a copy of the *Washington Star* that John had overlooked delivering on his nightly paper route), when he happened upon Mary and some of her girlfriends. Joe was in his signature, now white, denim outfit. Mary was rather cool in her greeting. When Joe walked away, one of Mary's friends was aghast.

"That's your dad?! I thought he was the Good Humor Man."

Joe should've been proud. He'd finally pulled off a success-ful disguise.

⤳

*JOE'S LACK OF MECHANICAL EXPERTISE KNEW NO BOUNDS.
Years later, when Joe was dying, Joe spoke of how much he appreciated my love and devotion, and asked if there was anything that he could do in return. My reply:*
   "Yes, Joe, would you mind learning how to play the record player?"

⤳

I DIDN'T KNOW HOW MUCH MORE HANDIWORK JOE COULD take, let alone how much our house could stand. He hadn't yet succeeded in hooking up with a specific country assignment at headquarters and was still working at a generic staff job. Maybe he needed the same thing my father had needed early on in his career: a Dale Carnegie course.

I broached the subject with Joe. He reacted as most men might at the idea of a self-improvement course: I was nuts!

But after some discussion with my father, Joe took the course and was elected president of his class. There I was, on his arm at his graduation banquet, reenacting the same role that I'd played years earlier when I'd accompanied my father to his Dale Carnegie graduation. Joe was a success—the Dale Carnegie Institute even asked him if he wanted to become an instructor in the evenings. Joe turned them down, but I think he was pleased.

He took his new skills with him to Langley, and started circulating his résumé with a vengeance, concentrating on Italy. (Maybe he was hoping he'd run into Fiorella.) I knew Italy was the posting he wanted, so it became what I wanted. I started using a little more garlic, imagined our farmhouse in the rolling hills of Tuscany and dragged out some opera records.

I recall being at home on Saint Patrick's Day during the time we were hoping he'd receive an overseas assignment. I was decked out in green, polishing up our hardwood floors to the sounds of Puccini, when Joe came in and lifted the needle off the record.

So much for Italy. It turned out that a friend of Joe's who

was of Italian descent and fluent in the language had gotten the
job. Joe was disappointed, but understood. He decided to concentrate his efforts on Latin America.

It was about time, I thought. The U.S. had been treating Lat-
ins like poor relations long enough—and, besides, I wanted to get
back there. Latin America was becoming the happening place. In
1959, Castro had come to power in Cuba and soon revealed his
Communist intentions, not long after we had welcomed him in
this country like a conquering hero.

One night Joe arrived home while I was in the basement doing
laundry. He called my name and, when I answered, he headed
down. He dispensed with the usual "hello" kiss and just plunked
himself down on the next-to-bottom step. His blazer, blue button-
down, and navy-and-red-dotted bow tie looked out of place in our
depressing, institution-gray laundry room. The laundry room had
nothing on Joe that night. There had been no takers on his résumé.

We talked. I had an idea. (No, not another self-improvement
course!)

"Joe, why don't you submit a color picture of me and the
kids, along with your résumé? They might think you have a meek
little Japanese wife and children."

Right about then I think Joe would've been glad to settle for
just that.

He stood up, straightened his flannels, shot me a long-suffering
glance and remarked:

"Bina, that's the most asinine idea you've come up with yet."

But I guess he figured he didn't have much to lose. He resub-
mitted his résumé, this time accompanied by a five-by-seven color
family photo.

Within a month, Joe began a six-month intensive course in
Portuguese. He'd need it for our next post: São Paulo, Brazil.

❧

I LOVE A GREAT SEND-OFF. BUT AT FIVE IN THE MORNING, IT
was, I thought, out of the question.

Our two cabs idled at the curb, as the Kiyonagas, busily silent, made their way out front with their assorted luggage. I took one last, fond glance out of our bedroom window.

At the curb were our close neighbors and even closer friends, the Gotts, kids included. All were beautifully turned out as if they were heading to afternoon tea. The irony of it was that they were better dressed than we were. Gene Gott had a silver tray weighted down with a pitcher of martinis and four frosted glasses.

Now that's what I call a send-off.

❧

CHAPTER XI

BEGIN THE BEGUINE

IF JAPAN TAUGHT ME TO BE A GOOD WIFE, BRAZIL TAUGHT ME
to enjoy it.

For us, Brazil was a dance on the beach, a knowing glance
across the room—all set to the gentle tug of a bossa nova, the
clink of ice, Joe's drag on a cigarette. For the first time in my life,
I didn't feel outrageous. I loved it. We loved it.

Sayonara, Tokyo. Good night, dear Washington. Bon dia, São
Paulo! What contrasts, what esprit! Yin and yang. Hot and sour.
I tend to equate a country in terms of taste. Japan enjoyed innate,
quiet good taste; Brazil's was flamboyant, but no less refined. If
Japan was the sublime choreography of a silent tea ceremony,
Brazil was a careening samba parade. (But watch your wallet.)

Everything in Brazil is larger than life. Their sunsets are surreal; their full moon is luminous to the point of being eerie; and
their Southern Cross on a starlit night outsparkles any possible
diamond concoction. The music makes you nostalgic, even if
you've never heard it before. The architecture, with its rakish
modernism, enjoys a verve and daring in tune with the country's
ambiance. And the people—everyone's having a good time and

it's contagious. I suspect that all Brazilians are born relaxed. You can see it in the way they walk.

They just don't seem to take life all that seriously. Maybe it's because so much there operates on faith—God is certainly not dead in Brazil. I could see as much as we flew into Rio. While Kamakura was presided over by the Daibutsu, a huge statue of Christ with outstretched arms reigns over Rio. In fact, they have a saying: "God is a Brazilian." The resulting philosophy is, to my mind, dead-on and inspiring: life is so short, and it's in God's hands anyway. So live it up. Brazil always opens another bottle of wine.

Whatever it was, Joe and I both immediately felt at home.

❦

BEING AN AMERICAN SPY IN BRAZIL IN 1960 WAS A LITTLE like being a Wall Street broker in the 1980s or a hippie at Woodstock: you didn't know what was going to happen next, but you knew you were in the right place at the right time. Joe was second in command at the CIA base in São Paulo. His cover: U.S. State Department official, a consul.

I was pleased. Daddy had been with the foreign service, and I liked the idea of State Department cover. It meant that Joe was now officially a diplomat, a real step up from "Department of the Army civilian." Diplomats worldwide are regarded as the cream of their local crop. A diplomat is automatically included on a country's "Diplomatic List," which gives you immediate acceptance within the country. Being on the list also gets you invited to important official functions (good for recruiting agents) as well as a lot of boring parties. You also are given diplomatic license plates and immunity, which helps, in a pinch, because the police will give you a wide berth.

The Brazil of 1960 was in political chaos—great for a spy. Jânio Quadros was president when we arrived. He was a dictatorial "independent;" he was also a little crazy and soon was ousted, creating a void. Joâo "Jango" Goulart, his vice president and a leftist, succeeded Quadros and started appointing avowed Communists to high-level government positions. Goulart didn't enjoy widespread support among Brazilians who were watching their economy erode. And he sure didn't have a lot of support in Washington: Goulart had come out against U.S.-imposed sanctions against Castro's Cuba. Washington would have shed no tears had Goulart taken an early retirement.

The Communists weren't wasting any time in Brazil, busy infiltrating their favored targets: trade unions, student groups, labor organizations. It was not a country the U.S. wanted to see go Communist—Brazil had the largest economy in Latin America and the sixth largest in the world. What's more, Brazil had—and has—unrivaled potential for future economic growth, i.e., the unique and vast resources of the Amazon Basin. (I don't doubt that a cure for cancer lurks somewhere in that boundless jungle.) I've always considered Brazil to be light-years ahead of the rest of Latin America and have a feeling its neighbors may never catch up.

Not only did Joe tend to be in the right place at the right time, he seemed to have the knack of dragging world problems along with him. In Japan, it was the Korean War and the China problem. By the time we hit Brazil, much of the Agency's efforts had shifted to Latin America because of Cuba. The Bay of Pigs fiasco (not entirely the Agency's fault, by the way) would come along in 1961, soon after our arrival in Brazil. Dark days for the Agency. The Cold War was getting a little too close for comfort as far as I was concerned.

Joe was tailor-made for Brazil. It made eminent good sense for the CIA to send a Catholic family man to Brazil, the world's largest Catholic country and a family-oriented one. It also made sense to send a Japanese-American operative to Brazil, which has a powerful Japanese Brazilian community. Today, Brazil has more Japanese than any country outside of Japan, and most of them are in the state of São Paulo. São Paulo even has its own Little Tokyo, "Libertade," where Joe could meet at tucked-away restaurants that didn't even boast a sign. When Joe recruited in that community, it was back to sashimi and sake.

❧

SÃO PAULO ITSELF WAS A SHOCK. I'D IMAGINED DETROIT SET to a Latin beat, but I found a sprawling, enthralling metropolis with Tokyo's size, New York's sophistication, Hong Kong's energy and Calcutta's shantytowns. I recall sampling the air as we drove in: a blend of hot diesel, overripe pineapples, and even a trace of barber-shop hair tonic. Interesting.

This was our first big move as a family of six. In time, I would perfect the drill, but, at the time, it seemed overwhelming. First, we had to rent our Leland Street house, which was traumatic enough. Next came the packing . . . and the shipping . . . and the storing. And school credits? They needed, of course, to be transferred for each child. Then the extensive physical exams began and, with them, a battery of immunization shots for the whole family. Not knowing what we'd be able to find in our new post meant that I had to shop two years in advance (the length of a normal tour) for clothes and graduated shoe sizes for all of the kids. I thought the advance shopping was the hard part, until the bills arrived.

And those are just the practical considerations of an overseas move; the personal ones were far more unsettling. Good-bye Gotts, good-bye Garfinckels, good-bye snippy nuns! Once we arrived in Brazil the process began all over again—in reverse. There was no ready sense of community, neighborhood or home. For a CIA family, home is wherever the family photo albums happen to be.

In fact, our first home in São Paulo was the Hotel Jaragua. I took a look around the hotel lobby and immediately zeroed in on the obvious. Brazilians may be born relaxed, but the women are also born beautiful. (With an occasional assist from the legendary plastic surgeon, Pitanguy.) I hardly recall the men.

One look and I knew I needed to revamp myself and my

wardrobe, especially if I was to gain acceptance into the circles
Joe was seeking to infiltrate (and keep Joe's eyes on me). The
worst of it was that I felt as if I'd just gotten off the boat—or
the plane. Let's put it this way: on the flight to São Paulo, I'd
worn a "Ladybug" dress made of black-and-white ticking with a
red elastic belt. (I'd ordered it through my old friend, *The New
Yorker*, for $12.95.) I noticed that Joe had taken a look around
that lobby, too, but I hid my irritation. You see, in Brazil you
don't get mad; you get manicured.

For the first time, I began to use some makeup, though little
jewelry, and I moved from bobs and bangs to a blunt, Sassoon-
style cut. I dyed my eyelashes coal black (risking blindness, it
turns out). I kept up a summer tan, and even managed to throw
in weekly massages at home. Taking advantage of a strong dollar,
I had my clothes tailor-made on Rua Augusta, São Paulo's Rodeo
Drive—everything from salmon linen shifts to black faille evening
suits (tuxedo-style with a slit floor-length skirt and small rhine-
stone buttons). A new Bina emerged.

❧

WE EVENTUALLY MOVED OUT OF THE HOTEL AND INTO A HOUSE
in one of São Paulo's *Jardims* (gardens), residential areas named
after various continents—*Europa* and *America*. Farther out of the
city were *Morumbi* and *Chácara Flora*—more rural but palatial.
We lived in one of the lesser *Jardims, Jardim Paulista*.

Our home was simple white stucco inside and out with black
grillwork and an enclosed *jardim de inverno* (winter garden) in
back. This is a peculiarly Brazilian invention. Unlike our gazebo,
it is an extension of the house proper and is totally enclosed by
garden. Whereas the Japanese garden had been a lesson in restraint, the Brazilian garden knows no bounds. It is a jungle in
miniature and just as varied.

In such sumptuous tropical surroundings, many an agent
would be recruited.

❧

A FEW MONTHS AFTER WE LANDED IN SÃO PAULO, I RECEIVED a cable—my father had suffered a heart attack. I called home (from a neighbor's phone; ours had yet to be installed) and spoke with my brother-in-law, Frank Cyman. Frank was a doctor and I respected his opinion.

"Frank, how serious is Daddy's heart attack?"

"Bina, it's pretty bad."

"Should I come home?"

Pause.

"If I were you, I would."

Joe was as upset as I was. He urged me to go, despite the cost, and the fact that we'd just moved in. I took a flight that evening for Washington and found my way to Arlington Hospital.

The hospital scene was grim. Daddy lay, barely conscious, covered by a plastic oxygen tent. Mother, Mary Ann and Frank stood by the bed. No words were necessary. Daddy was dying and we all knew it.

Later in the week, Daddy's condition worsened, his breathing became labored. It was hard to watch. Mother and Mary Ann were understandably upset, but their very apparent grief must have been doubly upsetting to Daddy—even in his semiconscious state. Frank and the attending nurse ushered them out of the room.

I approached his bed. I reached under the tent and grasped his hand. He'd been there for me my whole life. When the rest of the world had made me feel awkward, Daddy had always made me feel special.

Daddy certainly must have been aware enough to feel my touch, because he roused and smiled up at me:

"Gee, Bina, what I'd give for a cold beer."

Daddy died that evening.

❧

No "what ifs" or "if i'd onlys" between Daddy and me. We'd said and done it all when it mattered. But the ache was there. Joe felt it, too. He and Daddy had become good friends, and Joe had looked to my father for fatherly guidance. Joe had been given the choice of Brazil or Argentina. It was Daddy's suggestion that he go to Brazil.

"That," my father had advised, "is where the action is."

He was right. I came back to find an international crisis brewing. We were the only Americans in our neighborhood. The rest of the block was all Brazilian, except for a family from Lebanon nicknamed "the Turcos" by the neighbors. (Political correctness wasn't in yet.) It would be fair to say that John—very blond, very thin, very tall and very outspoken—wasn't well liked by the Turcos kids. They were relentless. John was brave, but outnumbered. He eventually turned to his big brother, David, who, in turn, rounded up a few of the neighborhood Brazilians and began a virtual soccer embargo against the Turcos kids. Joe had enrolled the boys in judo classes and had put up a heavy punching bag for them in the garden—facts not lost on the Turcos boys. David's strong-arm diplomacy worked, and within weeks, all of the neighborhood kids were drinking Todi together (the local brand of hot cocoa).

In the case of Japan, it's the Agency colleagues I remember; in Brazil, it is our Brazilian friends, especially our neighbors. I recall that Joe and I applied for membership in the Clube Harmonia, a small but tony club in São Paulo. Most of the consulate staffers belonged. We were the first to be turned down—on the basis of Joe's being Japanese. (I had thought we were past all that.) Then we appealed to our friends, "Peggy and Paul Brown," the deputy chief of mission, for help. They in turn turned us down—as members they didn't care to become involved.

It was our Brazilian neighbors who came to the rescue and secured honorary membership for us in the Clube Paulistano. The Paulistano had far better facilities, but we had not applied for membership because of the exorbitant cost involved. Now we were invited to be members, gratis. I recall only one other American family being members.

Country clubs weren't just some nice "perk;" they were a necessity for Joe's work. (Keep in mind how helpful that golf club membership had been in Tokyo.) At a club, you found a self-selected group of Brazil's elite relaxing along with their families. Access to business tycoons, government heavyweights, influential politicians and savvy journalists was just a tennis game away, to be followed by a poolside lunch while your kids played Marco Polo. The food was pretty good, too.

Interestingly enough, once word got out that we'd joined Paulistano our chief of mission, the consul general, applied for membership. No soap!

⸎

MEMORIES OF THE SUCCESSFUL JAPANESE LIBERAL DEMocratic party fresh in his mind, Joe tried to form a popular Brazilian Democratic Party, as a bulwark against the Communists. He sought the cooperation of the Christian Democratic Party, the Japanese community, and the Catholic Church. It never got off the ground. I'm not really surprised, given the ingredients. They were hardly a homogeneous combination. But the main resistance was Brazil's seeming predilection for military rule. The country was simply not ready for political experimentation.

But the Brazilians were *always* ready for a real party. In Japan I'd learned the art of the small dinner party, lavish in quiet detail. But Brazil was vastly different—if Joe was going to succeed at recruiting in Brazil, we needed to turn up the volume . . . and the treble . . . and the bass. I was happy to oblige.

Each party was approached with the meticulous planning of a military operation. Joe would first draw up a guest list. He wouldn't tell me why he wanted a Brazilian nuclear physicist, along with a telephone company CEO or political science professor; or whether some guests were already agents, potential recruits or merely sympathetic to U.S. interests. Sometimes Joe's purpose was to have others create relationships that he, in turn, could take advantage of. Having, say, an American businessman share a buffet line with a Central Bank vice president would work out to everyone's advantage. If Joe didn't recruit the Brazilian banker, he might recruit the U.S. businessman, grateful to Joe for a profitable—and informative—friendship with the banker. Another venue for intelligence-gathering.

Joe made it a point to include some people who were "in" politically, and some who were "out," on the theory that you never knew when an "out" might be "in" again. I think CIA was more farsighted than State in that way; in my experience, State Department parties tended to be heavy with current Establishment types, old "cliff-dwellers" no one had the courage to scratch off the invitation list. CIA tended to hedge its bets more, so that it would always find itself among friends no matter who was in power.

My role was to plan the festivities and expand Joe's core guest list to ensure some chemistry. Too many business types? Throw in a Brazilian Japanese abstract painter fresh from the Venice Biennale. Maybe an Amazon-based environmentalist to stir things up (all the better if he was good-looking, or had a charming wife). On the phone with a friend, I'd let drop the fact that the avant-garde playwright recently profiled in the *Jornal do Brasil* would be stopping by on his way to the airport. I always invited more people than we needed, figuring that a crowded party made for a better party. Joe would approve the final invitation list.

Then would come the fun part. I'd bring in the florist and our home would be transformed into an Italian garden, tropical isle or whatever the occasion demanded. I'd consult with the caterers to come up with a menu in the same vein. The afternoons were always a whirl of waiters bearing trays of Scottish salmon, beef carpaccio, or a papaya/mango/maracujá (passion fruit) assortment, weaving in through the trio of musicians setting up their instruments in the *jardim de inverno*. The music would continue the international theme with selections from Jobim, Jorge Ben, Sinatra, and the Lettermen.

The night of the party, Joe and I would dress while sharing a martini and listening to the band warm up. We had one rule: only tonic water with a twist of lemon once the guests arrived. Joe would brief me on each guest—their position, where they were schooled, their children.

When I think back to those nights, one image is of Joe giving his tie a final tightening tug before heading out to greet the guests in the foyer. The music would already be going, the maids in their starched, aqua linen uniforms standing demurely to one side with trays of *salgadinhos*. (I even made sure that they had their hair done.)

I've never been one to separate adults and children. We always included the kids at the parties, until their bedtimes. We were proud of them, and, besides, meeting our eclectic guests was part of their education. It's really something to see your ten-year-old holding forth in nimble Portuguese with an oil company exec, and especially disconcerting if you can't follow the conversation. Come to think of it, my kids always seemed more comfortable visiting with adults than kids their own age. Maybe it was an outgrowth of those parties.

Brazil's verve seemed to rub off on the kids—their preference was for the bossa nova over Motown. And, believe me, Brazilian fashion was not lost on them. Mary was sporting her new inch-high cork-soled sandals and the boys practically slept in their soccer shirts. To this day, it seems as though most of my kids spend half their time trying to get back to Brazil.

It was usually my role at our soirees to befriend a particular wife, whether she liked it or not. I would make it a point to visit about the family—any stray bit of information could sometimes help Joe clinch a recruiting effort. Ernesto Jr. wants very much to study urban planning in the States? Who knows, maybe a scholarship could be arranged. Securing scholarships to U.S. colleges for the children of important contacts was a favored CIA recruiting tool. (I imagine most universities have less dealings with CIA today. It made for bad press.)

My other image of Joe is of him clinking a glass at the height of the evening to offer a toast. His posture was casually correct, the toast usually just a brief word of welcome to our home. Nothing too studied or staged. When a couple would approach Joe to say they were leaving, he would always walk them to their car, even if it meant excusing himself from an important conversation. It wasn't just good manners, it was good recruiting. I think people were struck by Joe's thoughtfulness.

Many a party would be capped off with a final dance to "The Girl from Ipanema." Sometimes, it would be just Joe and me dancing after the guests had left, a weary band and the candle-lit banana trees the only spectators.

After the party—the next day, if it was a really good party— Joe would debrief me. Did the minister's wife say anything about her husband's upcoming trip to China? (No, Joe, but she did discuss at length *her* upcoming trip to Paris.) Did she mention why he had gone into the hospital last month? I never knew how the pieces of information I gave him ever fit into the puzzle of intelligence gathering—probably just another addition to the dossier. Sometimes I'd venture an insight into someone—whether I thought they seemed trustworthy, drank too much or talked too much. Joe would tuck the information away, without comment. He would never let on, but I could always tell if the night had been a success.

❧

YOU NEVER KNEW WHERE A CHANCE CONVERSATION AT ONE OF these dinner parties could lead. Through one of our dinner guests, I learned of Darcy Penteado.

The São Paulo Bienal, one of the world's foremost art shows, had just taken place, and Penteado, a portrait artist from São Paulo, had won first prize. My friend, "Flavia," took me to tea at the artist's apartment. I was overwhelmed. So many wonderful things in one apartment: a stand-up desk made of jacaranda wood from the Amazon, gold and silver foil flowers from a church altar in Northeast Brazil. Tea went well, and I reciprocated by inviting him for tea at our home.

Darcy Penteado pulled up to our gate at the appointed time. He drove a yellow Volkswagen Bug convertible and wore tan slacks, a silk shirt, and monk sandals—exactly as I'd expected a true artist would be dressed.

Just as we started toward the house, Ann wheeled up on her tricycle. She was wearing her pink-and-white-checked gingham dress with pink moire ribbons tied around her thick braids. She smiled at him with undistilled innocence. Darcy took one look at her, excused himself and went back to his car. Puzzled, I waited. He returned with a canvas under his arm. With apologies, he asked permission to sketch Ann. Permission? For one of the world's greatest portraitists to sketch my daughter?

Christmas morning, Darcy's black ink portrait of Ann was propped up against the mantel—my gift to Joe.

Ann's stardom was rather short-lived. Shortly after her "modeling" stint, she was invited on a local children's television show. The show's sponsor was *Pão Pullman*—a sweet sponge-cake-like bread (Brazil's answer to Twinkies). Ann, the only American child, was dolled up, proudly sporting her crisp school uniform. The children were giddy with wide-eyed excitement when the fabulous *Pão Pullman* was brought out. As the tray was passed, each child eagerly gobbled up their share. The tray came to Ann. This was her moment. Our moment. She politely demurred, and was *very* correct (always the diplomat's daughter) in her response:

"Thank you, no. I don't eat sweets."

❦

JOE AND I REALLY BEGAN TO FEEL LIKE A COUPLE—LIKE partners—in Brazil. Our "Joe and I" became a "we." Maybe it was the Communist threat, maybe it was my eyeliner. It didn't really matter—it was palpable. A friend of mine recently told me that she recalled that anytime Joe and I were at a party together, Joe would raise his glass to me from across the room and toast me with his eyes before taking his first sip. I remember those silent toasts, too. It was in Brazil that they started.

We probably have Brazil itself to thank—its enticing samba beat gets into your blood. It did for our marriage what a South China Sea cruise on a *Seabourn* does for a second honeymoon. There was just something about the place—we laughed more, danced more. We just *were* more.

For Joe, I think he found in Brazil a kindred spirit, a rustic sophistication that reminded him of Hawaii. It was the Wild West, with an air of adventure, seedy corruption (the best kind), and vast tracts of untamed *mato* (brushland). Brazil was what Molokai could have been with some imagination and decent music.

Japan had been too fragile; the States too developed; Brazil, by contrast, was primal, primeval, primordial. Joe reveled in it. He loved the rough burn and bitterness of a Bahian Suerdieck cigar; it exalted his sense of domain. He loved a Sunday afternoon at a friend's *fazenda*, or ranch. Some, in the interior of São Paulo State, were centuries old, with their own tiny, whitewashed family chapel. You'd look out from the veranda at the glistening horses stomping the earth. Farther off were groves of coffee and Persian limes. It was the Old World in the New World.

Even the foods were earthy. The charred meats of a Brazilian *churrasco* (barbeque) were unprocessed, gamey, salty. You could feast on the tender hump of a Brahman bull, a dozen other cuts of beef, broiled chicken hearts, alligator filets—all brought to your table impaled and sizzling on a rapier. The accompaniment was usually *palmito* (heart of palm stalks), arranged on the plate like toppled Roman pillars, or speckled, hard-boiled quail eggs.

And in Brazil Joe could return to his boyhood pastimes: hunting and fishing. He could fish for *pintado* (a grotesque and giant snorting catfish) on the prehistoric Araguaia River in the heart of Brazil, or venture out to its aptly named, piranha-infested tributary, the Rio dos Mortes (the "River of Death"). The bait for *pintado* was a severed piranha dangling from a steel hook. Spare piranhas were slow-boiled at the end of the day to create a fragrant broth and a reputed aphrodisiac.

Joe felt comfortable in Brazil. It helped that the Brazilians were often a unique racial blend. There were Germans and Italians in the south, Japanese in São Paulo State, descendants of African slaves in the northeast, indigenous peoples in the Amazon, and Portuguese everywhere. (One town along the Amazon River even boasted descendants of refugee U.S. Civil War Confederates, with names like O'Donnell and Sullivan.) The Japanese had intermingled with the Brazilians, creating, in the process, exotic looks with equally exotic, and sometimes comical, names: Winston Ichimura, Edson "Zico" Yamamoto. No more questions about his origins. A 6'4" Japanese who looked vaguely like Pancho Villa just didn't turn as many heads.

Let me qualify that: not as many male heads. I had always thought my enemy to be the ill-tailored Soviets bent on world domination, or Brazilian anarchists concocting schemes of revolution over sugar cane brandy. No, my enemy, it turns out, was the Brazilian Woman.

Don't get me wrong—those dinner parties were great. But every so often, one would be ruined, because I'd be too busy— busy watching Joe like a hawk as he conversed easily with some bejeweled femme fatale. Why did he have to wear that white Pima cloth shirt, set off so dramatically from his reddish-brown skin and ink-black mane? My husband did it for me. His sheer dynamism was the first thing that had hit me when I'd looked across that drawing room back at Michigan. I knew I wasn't the only woman to pick up on it. Wait a minute—what was that he just said about his upcoming trip to Rio? I was spying on a spy.

In all fairness, Joe wasn't much better—he'd barely let me dance with anyone before he'd cut in. The Brazilians don't tend to bother with any of this. They just go ahead and have affairs.

❧

MY NEW-FOUND BRAZILIAN ZEST MUST HAVE BEEN SHOWING when I attended a meeting of a brand-new organization of American government wives based in São Paulo—named (not surprisingly) the U.S. Government Wives Association. Some State Department and other non-agency wives had difficulty getting involved in the local scene because many of them didn't speak Portuguese. (I was lucky; I'd learned Spanish in Colombia, and CIA wives were offered language training before we left the States.) I made the mistake of making a few off-hand suggestions about possible activities and, before I knew it, was elected president.

I knew enough to be more wary than flattered—with four children at home, you don't want to have to spend your nights typing up newsletters and stuffing envelopes. But I thought it could provide Joe contacts, as well as rustle the timbers of a lot of wives who were busy complaining of what I considered their "cushy" lot. Charitable activities made sense, especially because none had existed connected to the Embassy before.

I would earmark Agency wives and try to match them up with Brazilian wives who might prove valuable contacts for the Agency. I placed one young Agency wife on the committee for the Clothes Bank for the Blind—the husband of the chairwoman of that particular charity happened to be a high-level congressman. It was for a good cause, and these kinds of social associations really paid off.

❧

THANK GOD I TURNED TO CHARITABLE WORKS IN BRAZIL. Through the Clothes Bank for the Blind, I had a chance to do something really needed and useful. And to meet Helen Keller.

Helen Keller was one of the most admirable women of the twentieth century. Despite being deaf, mute and blind, she obtained an undergraduate degree from Radcliffe. I was privileged to meet her, and her traveling companion/interpreter, when they visited the Foundation for the Blind in São Paulo.

When I was presented to her, she reached out her hands and clasped each side of my face to trace its contours and felt my voicebox as I spoke. Then Miss Keller turned to her interpreter and tapped out a message in her hand, something to the effect of thanking me for my efforts on behalf of the blind.

I truly felt I was in the presence of the most remarkable person I'd ever met. Years later, I would have the same feeling when I met Mother Teresa.

Another famous visitor was Helen Hayes, in São Paulo on tour in *The Glass Menagerie*. I was asked to be her hostess during her three-day stay. Me, a drama major, squiring the "Queen of the American Theater." Not my idea of heavy duty.

Helen Hayes was a delightful, unassuming person. We had a fine time visiting various Broadway friends of hers who had sought exile in out-of-the-way colonial coffee plantations.

On the last evening, she starred in *The Glass Menagerie* (and brought down the house). An after-theater supper had been planned, and Joe and Helen Hayes met for the first time. As an approving aside to me, she told me I was fortunate to have such an attentive and charming husband, and that we should enjoy our time together. She related a story about her late husband, playwright Charles MacArthur.

Early on, when Helen had been an aspiring actress and Charlie a struggling playwright, they had met at an off-Broadway party. When Charlie spotted her across the room, he didn't waste any time. By way of introduction, he offered her some peanuts from a paper bag he was holding, with the words:

"I wish they were emeralds."

Years later, when they were married, and both were world-renowned (he'd coauthored *The Front Page* with Ben Hecht), Charlie made a trip to India, despite his terminal illness. His wife was at the airport when he alighted from the plane. He handed her a small paper bag—filled with emeralds—and said:

"Helen, I wish they were peanuts."

❧

HERE I WAS, SERVING ON CHARITY BOARDS AND ATTENDING ladies' lunches—filling a woman's conventional role—while, at the same time, I was turning the fruits of my labors to the most unconventional of ends—helping my husband spy. That's what you do as a CIA wife—you keep your eyes and ears open, even at something as mundane as a school tea.

The kids attended Chapel School. It was there that I struck up a conversation with "Cecilia." It just so happened that her husband represented a major U.S. movie studio in Brazil. That sounded promising.

"*An American in Paris* is playing. Let's get together and go," I suggested. "Why don't you come over for lunch Saturday? Bring 'Edward' and we can make it a double date."

You get the idea. She was soon on her way over for lunch with Edward, her husband. We started out with *caipirinhas*— cocktails made with a potent Brazilian cane whiskey, mixed with limes and sugar. It sounds terrible, but it tasted great. Joe conducted a lot of business on the strength of them.

After several rounds, Joe and Edward retired to a quiet corner of our *jardim de inverno*. They appeared deep in conversation and, not wanting to interrupt them, I had Aparecida, our maid, serve them out there. (Half of your job as a CIA wife was knowing when not to interrupt.) We were having *feijoada*, the traditional Brazilian dish, served each Saturday: a rich stew-type meal made of every conceivable part of the pig—tail, ears, entrails—all thrown into a pot and simmered for hours. We did it up in true Brazilian style, with an accompaniment of black beans and rice, *farofa* (a sawdustlike meal made from manioc root), sliced oranges and shredded mustard greens. A meal of *feijoada* made for a long afternoon, ideal for an in-depth discussion.

Later, the two men joined us for dessert and *cafezinho* (Brazilian espresso). I learned later that Joe had recruited the studio exec that very afternoon.

❧

JOE VIEWED EVERYONE AS A POTENTIAL AGENT. AS IN JAPAN, Joe's strong point continued to be recruiting political figures and journalists. Getting to them required careful groundwork—you don't just pick up the phone and invite them to a dinner party at your house. Joe maintained a notebook, which he kept in his office safe, for each country in which we were posted. For every man or woman he was targeting, he had a picture, name, address, phone number, and as much biographical information as possible, including information on the spouse and children. There were usually extensive notes and observations on the targets ("Most often seen accompanied by . . .") and favorite haunts ("Frequents the dining room at . . ."). Even though I never saw it, Joe's notebook had a tremendous bearing on our lives since I often chose my friends at his direction. I sometimes discarded them at his direction, too; a nosy friend was to be avoided.

Joe had many valuable contacts within the American business community. Some were so willing to cooperate that they actually helped to recruit agents within their employ. A few occasionally served as "cut-outs," that is, they acted as a go-between in meetings with agents, so that the agent would not be seen in direct contact with the CIA operative. Sometimes, they did "dead-drops," leaving material in a prearranged spot to be picked up by the designated contact. Joe was wary of these operations because of the risks involved and the possibility of exposing a high-level businessman.

These Americans rarely accepted compensation; they volunteered their services. Their interest was twofold: protecting their business holdings in Brazil (a big consideration) and helping the U.S. government.

Sometimes CIA operatives would pose as U.S. "businessmen." In fact, some of the agency's most successful operatives were "businessmen." Joe worked closely with one such operative. This man served as a perfect example of what a CIA undercover, or "deep-cover," business operative should be. "Mike" learned to speak Portuguese fluently, immersed himself in things Brazilian, even began driving a motorcycle rather than a car. He built a successful business from scratch. Like all good salesmen, he proved to be better at selling his own product than he was at recruiting agents. Nonetheless, despite his crowded schedule of business transactions, parties, and golf matches, he managed to recruit his share.

Mike focused on the wide world of sports. He was especially effective at recruiting Brazilian soccer stars (world travelers themselves) and members of visiting Iron Curtain sports teams.

Mike was also adept at dealing with a long-standing White Russian agent who was highly placed in the largely European society of São Paulo. The Russian was a confirmed anti-Communist. He headed his own enterprise that entailed a lot of travel to Europe and various Latin countries. In essence, the agent's mission was to sell these countries and Brazil on the idea of political activism. He did so by using an academic operation, aimed at introducing democratic tenets into school programs and propagandizing these ideals through the use of seminars and public debates—all orchestrated by him.

❧

SCHOOL TEAS, SOCCER PLAYERS, WHITE RUSSIANS—WHEN you're a CIA operative, you use every available means to meet people, recruit agents, and garner information. Even your Church.

The Catholic Church has always worked hand-in-glove with CIA in combating Communism. The Church not only took a strong anti-Communist stand publicly, it also operated very efficiently beneath the surface. The relationship with the Church would reach a whole new level in the 1980s when then-CIA director William Casey, according to published reports, met secretly with Pope John Paul II to forge an alliance to bring down the Soviets.

In fact, one of Joe's best agents was a priest who was a federal deputy (equivalent to a congressman in our House of Representatives) and later became a senator. There were about sixteen political parties in Brazil when we lived there in the sixties. Father "Marco" was invaluable in keeping Joe apprised of what was happening in Brazil's Congress. He also engineered some congressional bills, dealing mainly with the critical issue of relations with Cuba. Despite his political position, Father accepted modest remuneration. He lived in impoverished circumstances, and was the sole support of his parents.

As a Catholic wife and mother, I had my own ready—and sincere—entrée to priests: I was concerned with my childrens' parochial schooling.

At one of our son David's soccer matches, I met another priest. Father "Ricardo" was tall, attractive—and lonely. I engaged him in a long conversation about the state of Catholic education in Brazil. He quickly accepted my invitation to dinner. That's when I introduced him to Joe; I was relieved when he shifted his attention from me to Joe and their common causes. I know Father Ricardo became an invaluable agent, though I'm not aware of what services he performed.

Joe might have taken the whole collaboration with the Church a little far. He seemed to spend half his professional life in church. He was pleased, and proud, when David was selected as an altar boy to assist the priest at Mass at the chapel on Alameda Franca. The Mass was at 7:00 A.M. Though it was early, Joe insisted on driving David. But, often, instead of going into church to hear Mass, Joe would wait in the car. "Tomas," the other altar boy, had a father who waited outside, too. It develops that Tomas's father was an agent of Joe's. So much for religion!

The confessional in the chapel on Alameda Franca was especially busy. Joe met there regularly with one agent, touted to be chief confidant to the head of São Paulo's Communist Party, the largest in Brazil. The man operated as a "Deep Throat" of sorts—Joe never knew his name.

Arrangements for their initial meeting were cloaked in secrecy, the most elaborate precautions Joe ever experienced. The agent-to-be felt he was being watched and was running scared. Joe and the agent would meet in the confessional, arriving and leaving the confessional at different times so that they would not be seen together. Joe would enter and kneel, ostensibly to confess, and open the meeting with the traditional "Bless me, Father." If, by

some chance, a priest—instead of the agent—was in the confessional, Joe would continue his confession. Joe got something either way—information or absolution.

Joe never actually saw the agent's face, which was probably a
good thing—word had it that the man was badly scarred, having
been subjected to torture somewhere along the line. The agent
provided substantial information concerning basic Party policy,
most of it already known to the base. At least it established his
bona fides. As time went on, his demand for money increased as
his output decreased. He had a large family, many debts and a
mistress. On occasion, his information proved faulty. Either the
agent was withholding knowledge, in hopes of more pay, or he
honestly didn't know.

Joe called for a showdown. He stipulated that there would be
no increase in salary unless the quality of information improved.
The agent reacted in kind: arriving late at meetings, refusing to
reduce information to paper. Finally, during a meeting with one
of Joe's subordinates, the agent threatened Joe's life—and that of
our family. The threats weren't idle; the man was known to carry
a gun, and his connections within the Party were rock solid.

The next morning, I opened the curtains to see a guard with
a submachine gun slung over his shoulder pacing in front of our
house. You can imagine how I reacted. Joe calmly explained away
the guard's presence, claiming an "imminent revolution." Meanwhile, Joe told the driver and school authorities to be on the
alert. He didn't tell me of the real threat until years later.

The guards (they changed every twelve hours) almost proved
to be a luxury that we could ill afford. They ate more than their
share of *feijoada* and monopolized the help. The maids spent most
of their time entertaining them.

Shortly before we left Brazil, we were at a party. Joe heard a
strangely familiar voice. It didn't take him long to place it—the
voice from the confessional. Joe was astounded to realize that all
this time he'd been dealing with, and threatened by, a prominent
Brazilian business figure.

⁓

I MARVEL THAT JOE HAD TIME FOR SO MUCH RECRUITING. LIFE within his office had to be more intriguing than any happenings outside. One of his colleagues was bedding down with his secretary. Joe knew. He found out the hard way. One night he stopped by the office to send off a cable. At last he realized why his coworker had installed a sofabed in his office.

To complicate matters, Joe's chief case officer lost his wife to an undercover agent. Another case officer divorced his wife for a secretary, and yet another operative was psychosomatic and refused to work.

The last thing you'd think he needed was more confusion at home.

⁓

I FIRST RAN INTO MIGUEL, OUR PARROT, ON THE BEACH AT Guarujá. He was the barker, or entertainment, for a beachfront *lanchonete* (snack bar). He was rainbow colored and colorful. He'd whistle at passing bikini-clad beauties, and they'd glance at him in surprise:

"Look at that! A whistling parrot!"

He'd reward them with a dance, first a twirl on one leg and then a twirl on the other—to the beat of a ribald tune.

Hooked, they'd stop for shrimp on a stick, a Brahma Beer, or a T-shirt.

I'm not the parrot type, but that bird was a scream. I had to have him. I approached the proprietor, busy making change for a *misto quente* (grilled ham and cheese) sandwich.

"What's the parrot's name?"

"Miguel."

"I love him. Would you consider selling him?"

"Sell him?!"

Shock. Dismay.

"Why, it would be like tearing out my heart. He's a member of the family."

I mentioned a possible price, hoping it would still his heart. Nothing doing. Miguel was indispensable for business.

But more. They were so close that they shared the same bed. I was feeling like a home-wrecker.

This situation called for Joe.

A few minutes later we were back, and Joe closed the deal. The proprietor looked less than sad as he counted his cruzeiros.

"Bina, what, in God's name, are we going to do with a parrot?"

Joe needn't have worried. Miguel would soon have company.

She came in the form of an Easter present. A priest/agent of Joe's had been given a wonderful black greyhound puppy by a parishioner. Father "Joaquim" lacked the space in the parish house to keep it. So, on Easter morning, he magnanimously left the puppy leashed to the gate outside the Church with a big bow around her neck. Every child at Mass admired the dog—but it was Joe who got the prize! After Mass, he coolly untied the leash and walked the frolicking pup to our waiting car. Maggie, the greyhound, had joined our family.

❧

"JOE, MAGGIE'S STILL RUNNING BEHIND US."

"Don't worry, Bina."

"But she's gaining on us." (She was a greyhound, after all.)

"She'll give up at the end of the block."

Joe and I were gunning down our street, Rua Conselheiro Zacarias. Maggie's chasing after the car had become something of a ritual. She hated to feel left out.

We were off to dinner. Middle of the week, no kids, sky's the limit. I had no idea why. All I knew was that Joe wanted me to dress for the occasion. Delighted, I got out my beige faille dress, beige silk slippers and prized Japanese pearls. Ten millimeters in size, perfectly matched, they were my favorite gift from Joe. In my excitement, I dropped them on the tile bathroom floor and cracked one of the pearls. This was proving to be a pretty expensive evening already.

We started off at the Prince Hotel where we downed two martinis each and tuna sashimi (in homage to our evenings at the Imperial in Tokyo). From there, it was on to Baiuca, for steak tartare and Joe's favorite wine, Châteauneuf-du-Pape. We finished off the meal with Brie and mangoes. Our last stop was the Ca' d'Oro Hotel for cognac and *cafezinho*.

Joe smoked the entire time. Over my *cafezinho*, I watched his ritual. He'd tap out the cigarette, usually a Pall Mall. A furrow of concentration, cigarette dangling casually. The whir of the match and sulfur flash. The orange glow and brief inhale. Then he'd lean back unhurried, midst the blue haze, suffusing a gentle satisfaction. It always got me.

At midnight, we were in the car heading home. Joe was smoking. He took a look at his watch, lowered the window and pitched out his lighted cigarette. He told me his news.

"I have an ulcer. It's pretty serious, Bina."

I should have known.

It had been too much. All the ingredients had been there: the threat of a Communist takeover; the pressure of recruiting agents, who trusted Joe with their lives; the double agents who threatened *our* lives; the armed guards round-the-clock; the morale problems at the office—plus an overly busy household.

❧

AND ALWAYS PRESENT WERE THE TWO OPERATIVES WHO HAD failed to return from a long-ago secret mission. Joe learned that Jack Downey and Dick Fecteau were still in solitary confinement. Jack's mother refused to be discouraged. She made frequent visits to Washington trying to secure his release. Joe would get periodic updates from contacts at headquarters.

Their situation may have been grim, but the two men still

managed to maintain a sense of humor. Word had it that the Chinese had held a mock trial for the men a few years after their capture. It was the first time the two had seen each other, having both been in solitary. Dick and Jack were both clothed in that one-size-fits-all Mao-type commune garb. Dick glanced at Jack and muttered,

"Love your tailor."

∾

THE ULCER DID MAKE JOE REST, AND HE NEEDED IT. HE HAD to give up everything he liked—smoking, drinking, and coffee. He could only have white foods, such as rice, potatoes, creamed chicken, creamed soups, custards and milk. He was to have absolute quiet—not easy with our home menagerie underfoot.

Joe was always difficult; when giving up smoking, he was unbearable. Mealtimes were especially traumatic. The children could do nothing right. John talked too much, Ann mumbled, David had his elbows on the table. Joe was really suffering, and we were suffering right along with him. I reached the point where I actually begged him to take up smoking again.

We weren't alone. Aparecida, our maid for the three and a half years in Brazil, confronted me one evening in the living room. She was dressed in street clothes and had obviously been crying. I was astounded. She was an exceptional person (I draw the line at calling her my best friend), and we'd never had a cross word. She was having trouble controlling her voice.

"Dona Bina, I'm leaving."

Before I had a chance to answer she went on:

"I've been happy here. Your family is my family."

By now I was practically in tears, too.

"But I don't think I can stand having Senhor not smoke anymore."

∾

To make matters worse, I'd just gotten out of the hospital after suffering another miscarriage. The doctors had told me that I would need extensive surgery to carry another child. Being thirty-eight years old didn't help. None of the doctors was willing to risk the surgery. I finally decided to put the whole issue of pregnancy behind me.

That's when my "avenging angel" friend, Angelina, came to call and found me in the garden.

"Bina, I was so sorry to hear you've been ill."

"Oh that's all right—it was worth a try. Joe feels four is enough of a family and why push our luck."

Angelina stood there and stared at me, before stating the obvious.

"But Bina, you have to have a Paulista."

She was right. Of course we had to have a Brazilian! Brazil had given us so much. We had to give something back.

I'd heard of a great German doctor who was more imaginative than his peers, and more daring. I resolved to see him. He agreed to perform the surgery.

Joe was opposed—it was major surgery. But he realized how much it meant to me, and he hated to turn me down. My decision made, he was unwavering in his support.

I came out of surgery in October of 1962 to find that the world was on the brink of a nuclear holocaust—the Cuban missile crisis. Joe worked round the clock and visited me when he could.

It was a greatly relieved Joe who turned up at the hospital one evening. Kennedy had called Khrushchev's bluff. The catastrophe had been averted, at least for the moment.

Two months later I was pregnant.

❧

Brazil was a popular destination for American dignitaries in those days. Robert Kennedy visited Goulart to urge more pro-U.S. policies. Vernon Walters, a special headquarters envoy, reviewed the volatile situation. Rumors of a coup were rife.

By the end of 1962, Goulart's government was in real trouble. People had taken to the streets; they were probably some of the best-dressed protesters on record. Even when it comes to fomenting revolution, the Brazilians had unusual flair.

Society women, including some of my luncheon friends, marched to protest Goulart's leftist government. They were resplendent in mink and pearls, with Mercedeses and chauffeurs at the ready. The parade was routed on Avenue Paulista, near our street. I walked to the corner and watched in fascination as the matriarchs of São Paulo strode by in dignified silence. I spotted two neighbors and waved. Goulart was too leftist for their taste—and pocketbooks. I still find it interesting that those captivating Brazilian women were more politically attuned, and active, than their North American counterparts.

Their effort found support in another venue. At close of business one Friday, an unidentified Brazilian turned up at the U.S. Consulate General. He hinted at the nature of his business to the receptionist. She, in turn, turned him over to the United States Information Service (USIS) man. The visitor stated his case. He was an ex-military man and a concerned Brazilian. He was alarmed at what was happening to the economy and to the country. He claimed to be the head of a group of business and military men working clandestinely toward the overthrow of "Jango" Goulart.

The USIS man called Joe. Thus began Joe's largest operation in Brazil.

Joe began meeting with the group. Brazilians are great talkers, but it soon became obvious to Joe that someone other than the group present was masterminding the whole undertaking. Joe confronted them and demanded to know who the actual head of the organization was. They conceded that Joe's suspicions were well founded, and in a dramatic manner, Joe's contact withdrew his calling card and wrote a name on it—Gen. Olympio Mourão Filho Jr.—a prominent general. (The fact that the contact was related by marriage to Mourao came out only weeks later.) The group had hoped to withhold this knowledge from Joe—in the intelligence game, the fewer people privy to information the better.

At Joe's insistence, an introduction was arranged with the general. The meeting was set for 3:10 A.M. in a park where Mourão walked his two bull mastiffs. Joe and he couldn't mistake each other. The general stood a trim 5'1" and Joe 6'4".

They met unaccompanied, with only a recognition signal as their guide. Joe was to offer Mourão a cigarette, with the apology, "I hope you don't mind smoking a Marlboro." Whereupon the general was to declare, "Thank you, I'd rather smoke my own, but I'll be glad to join you in a smoke."

A natural affinity developed between the two men. Both were infantrymen and both of humble origins. They spoke the same language.

The general requested money and arms, especially light artillery. Joe promised to make a note of it, but was secretly convinced that providing arms to the military, in this case, was unnecessary. What they needed was money and an American seal of approval for their undertaking. Once the coup was an accomplished fact, they had to be able to count on U.S. recognition and support. Mourão hardly needed advice on things military, but he lacked the political know-how that Joe could provide. Joe was to be his conduit and confidant. He also helped the general perfect his intelligence organization.

I'm assuming Joe must have discussed all of this with his superiors and gotten the go-ahead to work with Mourão. (I thought Joe was pretty good, but even I knew he had his limitations. He could hardly have engineered the overthrow of the government of a country of almost 100 million by himself.) I think Vernon Walters was in the picture. I recall his being at our house.

The general was being watched, as was Joe. It was imperative that their acquaintance assume the semblance of friendship. As the first step, Mourão invited us to dinner at his home to meet his wife, Maria. I knew nothing of Joe and Mourão's plans, just that Joe needed an innocent ruse for all of us to get together.

During the course of the evening we all hit upon a scheme.
The actual discussion took place in the garden for fear of possible
bugging of the house. Just as the general and Joe were much
alike, Maria and I were totally dissimilar. She was beautifully
coifed, while my style tended to be understated. Valor was the
better part of discretion, it seemed, as Maria suggested:

"Bina, how would you like to be transformed?"

I figured it might be fun. I agreed.

With that Maria went into the house to return later with what
appeared to be an average-size suitcase. I thought she was going
to completely outfit me. "This case contains the key to our future
success," she declared. Maria opened the case to display a huge
cosmetic collection—false eyelashes, wigs. The works.

It was decided. I would be the guinea pig. Twice a month,
the four of us would get together so that I could be transformed.
(Interspersed were occasional visits to beauty shops, without husbands, to add to the authenticity of our friendship.)

And so it was that while Maria was curling my lashes, the
general and Joe were plotting a coup.

❧

APARECIDA WAS EXPECTING A BABY AT THE SAME TIME THAT I
was. She told me that she was pregnant when she first realized
it. She was unmarried and wanted to have an abortion.

I told her that, should she have an abortion, she could never
set foot in our house again; should she choose to have the baby,
our home would be hers as long as she liked.

Aparecida stayed but she was concerned. She sensed that the
prospect of a pregnant maid would be unsettling for the whole
family.

Little did she know. The sight of Aparecida, large with child,
answering the door and conducting guests into the living room
where I, equally large with child, would receive them, led to some
chiding of Joe. The remarks of friends ranged from, "Hmmm, I see
you've been hard at it" to, "Nothing like keeping it in the family."

Joe was a good sport about it—that is, until we were awakened at 3:00 A.M. one night by frantic knocking on our bedroom door. Zelda, our cook (and Aparecida's sister), was calling me.

"Dona Bina, Aparecida has to go to the hospital. The baby's coming."

I nudged Joe. He was a sound sleeper; otherwise how could he have failed to hear Zelda?

"Joe. Wake up. You have to take Aparecida to the hospital."

No response.

"Come on, Joe, we've got to help her."

Joe turned over and faced me. He was awake all right, but he was having none of it.

"Bina, are you crazy? Do you think I'm going to turn up at that hospital with our maid?"

(I guess this was the last straw.)

"If you're so anxious to help, you drive her."

"Joe, you know I can't drive."

(I was about to deliver myself.)

Joe must have realized that it was pointless to argue, so we compromised. He called a cab. I watched as the two sisters walked out to the cab in the dark. It made me sad. Zelda was carrying the suitcase that Aparecida and I had prepared. It held a baby's complete layette—mostly leftovers from my own baby's. Aparecida looked so alone and so scared.

She was delivered of a baby son, João, at six-thirty that morning. The baby was placed in the care of Aparecida's mother. Two months later, Aparecida left for Washington with us. A few years later, she married an American, a Treasury Department employee. On their honeymoon, they traveled to Brazil—to pick up João.

❧

TWO WEEKS LATER, IT WAS MY TURN.

On my thirty-ninth birthday, September 18, 1963, I went into the hospital to deliver our Paulista.

Brazilian hospitals don't supply baby clothes or blankets. You bring your own. The girls and I had spent months assembling the baby's wardrobe. I'd had sweaters, booties, leggings and blankets knit at a shop on Rua Pamplona. All sported matching satin ribbons. I'd had *batiste* shirts, gowns and blanket covers embroidered at a boutique on the Rua Angelica. I'd hemstitched the receiving blankets. (We were all looking forward to this baby.) My seamstress made the bassinet skirt—white dotted Swiss with cerulean silk lining. Even the mattress had to be specially made because the Brazilian version was too lumpy. Joe wanted something firm ("to support the baby's spine").

Particular attention had to be paid to the first outfit worn by the newborn. It was to be yellow—a lucky color in Brazil.

Paul Yoshio Kiyonaga was born at 1:00 A.M. on September 19, 1963—one hour after my birthday. It was only fair. He deserved a birthday of his own.

Brazil was my favorite place to have a baby. The doctor and nursing staff at the Seventh-Day Adventist Hospital in São Paulo were competent. More important, they were interested. They never lost sight of the human element; they even managed to turn baby-birthing into an enjoyable experience.

I received no anesthetic (I really didn't feel the need for it), and Joe sat with me until I was wheeled to the delivery room. Actually, he helped to wheel me. Always courteous, the doctor invited Joe inside for the big event. Joe declined. Five minutes later Paul was born. He was not the best-looking baby that Joe and I produced, but he was far and away the best dressed.

Within ten minutes of Paul's birth I was holding him in my arms. He was a vision in yellow—from the tips of his yellow-bootied feet to the yellow batiste ruching around his neck, not to mention the yellow sweater and leggings. I can't figure out why the nurses failed to cap him with the yellow bonnet. They were Brazilians. Their restraint surprised me. I mean, that boy was ready for a party!

It's a good thing. Joe, Paul and I were no sooner settled in my room than the doctor arrived, chilled champagne and glasses in hand.

I thought the champagne was celebration enough, but the

morning after Paul was born, Joe turned up surprisingly early. I guess he couldn't wait. He kissed me good morning and pulled a small black velvet box out of his jacket pocket. He opened it and out of its white satin interior he removed a gold ring. He smiled as he slipped it on my finger and said:

"Thank you for Paul."

The gold band held two small gold roses with a blue-white diamond in the center of each. He had designed the ring himself and commissioned Maria Mourão's jeweler to craft it. Twenty-nine years later, Paul would give it to his wife, Debbie, as her engagement ring.

❧

SIX WEEKS LATER, JOE'S TOUR OF DUTY IN BRAZIL ENDED. WE sailed from Brazil on the USS *Moore McCormack*. It was our private Noah's ark—each of the kids plodding up the gangplank, two by two, followed by Maggie and Miguel, with Aparecida, cradling Paul, bringing up the rear.

Our spirited send-off was on Joe's birthday, Halloween. Our Brazilian friends turned out dressed in costumes for the occasion. Have you ever seen a Brazilian Halloween? Well, don't bother, neither had they. But they managed to turn our send-off into a real "Carnaval." They even brought along their own samba band. It was so lively that some of our friends nearly missed the boat—back to shore, that is.

❧

EVENTUALLY, WORD OF GEN. MOURÃO'S POLITICAL ACTIVITY reached Goulart's ears. Fearful of the general's growing power, Goulart transferred him to Juiz de Fora, a military nowheresville located north of the capital, Rio.

On April 1, 1964—April Fool's Day (and six months after we left Brazil)—General Mourão and the First Army marched on Rio and overthrew the government in a bloodless coup. The headlines the next day declared it one of the biggest events in Latin American history. The U.S. government quickly conferred its blessing.

Brazil would live under a military dictatorship for the next twenty-one years.

⊱

CHAPTER XII

KEEPING THE FAITH

WE DOCKED IN NEW YORK. OUR PLAN WAS TO SHOW THE KIDS the "Big Apple"—trek up to the Empire State Building, bask in the neon lights of Broadway. We overestimated our mobility. Everything had to be arranged in duplicate: taxis, dining tables, etc. Instead of spending the prearranged three days in New York sightseeing, we took off for Washington within four hours. We just couldn't cope.

Washington was little better. All hotels displayed signs forbidding pets. Never mind. One look at our dispirited party emerging from two Checker cabs was enough to influence their hearts, if not their cash registers. Maggie, the greyhound, howled all night. Miguel, the parrot, flew away only to be found two days later. Aparecida, the maid, sobbed until dawn. She'd given her fluffy pink winter coat (purchased by me) to her sister as a farewell gift. Only the Kiyonagas slept.

Our problems were compounded by the fact that we'd lost touch with our luggage somewhere between New York and Washington. We were all cold. It was November 15, 1963, and we'd arrived just in time for winter.

Joe and I were looking forward to being in Washington during

the Kennedy era. We wanted to be part of the "New Frontier."
We wanted the kids to know this president who could stare down
the Russians while enjoying a Cuban cigar. So far, we'd seen the
excitement only from a distance.

I think Joe felt a little bad about whisking the kids out of
New York so quickly. So he made them a bigger, better promise.
Now that we were back in Washington, the first chance he got,
he was going to take them to see the president.

❧

IT WAS A FRIDAY. WE'D BEEN BACK ABOUT A WEEK AND RENTED
an apartment at Alban Towers in Washington, while our house
on Leland Street was being refurbished following our rentor's departure. The kids were just settling into their new schools. Joe
was back at Langley. And I was nursing Paul, enjoying the solitude
of a weekday, contemplating what color to paint the nursery.

That's when Aparecida came running in.

"Dona Bina, something terrible is happening."

I scooped up Paul and followed her back to her bedroom. I
arrived just as Walter Cronkite was removing his black-rimmed
glasses. So it was there, on Aparecida's tiny black-and-white television that I heard the news:

"President Kennedy died at 1:30 P.M. central standard time
today."

❧

*JUST A FEW YEARS EARLIER, JOE AND I HAD FILED INTO THE
USIS auditorium in São Paulo to watch the Kennedy/Nixon debate,
the first televised presidential debate. I can still see John Kennedy.
Fresh from Palm Springs—tanned, toned, and toothy—he was standing next to what appeared to be a perspiring ghost. It really didn't
seem to matter what they said.*

*Later, the consul who notarized the absentee ballots told us that
we were the only couple in the consulate who had voted for Kennedy.*

If he'd had his way, he would have "thrown our votes out the window."

❦

JOE WANTED TO KEEP HIS PROMISE. OUR CHILDREN WEREN'T going to stay at home "while history happened." He took Mary, John, David and Ann to the viewing at the Capitol and then to the funeral procession along Pennsylvania Avenue. Paul and I camped out in the living room in front of the television.

If you lived through that time, it's not the conspiracy theories that plague you. It's what was lost that day in Dallas.

The media can say whatever they want about John Kennedy now, but they can't change how we felt about him then. For a brief moment, it wasn't politics as usual. It was personal.

As for the Kennedys, of course we idealized them. We thought we knew them, their style, their summers at Hyannis Port. But in those moments following her husband's assassination, Jacqueline Kennedy proved we didn't know the first thing about her. Mourning illuminated a side of her we'd never seen. It took a lot more than good taste to stand in that blood-stained pink Halston suit on Air Force One as Vice President Johnson was sworn into office.

President Johnson was handed the political reins, but the soul of our country remained in her hands. We watched her for our cue. She was the closest thing to royalty we'd ever known. There she presided as widowed queen. Her dignity calmed us. Her bearing emboldened us. With a single public tear, she could've brought this country to its knees. But she didn't.

Because of her, we were intact. Although never the same, we'd be all right. Our country, through *its* tears, breathed a collective sigh of relief.

I can't thank her enough.

I remember wondering how she could deal with it. A widow at her age. And her children. So young. They'd be hearing about their father the rest of their lives without ever really knowing him.

❦

I DON'T SEE HOW ANYONE CAN DEAL WITH TRAGEDY WITHOUT faith. For me, faith means seeing God in everything. (As some saint said: "God is in the pots and pans.") That everpresence of God inspires you—even obliges you—to return that smile, plant a seed (or pick a flower), love your neighbor—as though you were dealing with God himself.

God's always been big with me. I talk to him constantly. If someone on the road swerves in front of me, and, by slamming on my brakes, I've barely avoided a major accident, I'll find myself whispering, "Thank you, dear Lord." If God is big with me, His Mother is my dearest friend. I visit with her; talk things over with her; and actually dump things in her lap if they prove too much for me. Teresa of Ávila said, "God alone suffices." I have an addendum: "And His Mother, too."

I'd like to think that it's mainly love of God that makes me devout. But I doubt it. I'm pretty sure fear of hell has a lot to do with it. Those years of Catholic schooling had their effect.

Joe came from a different school, so to speak. Before his late-night sessions with Father Roggendorf, Joe had been a nominal Catholic. His lack of religious schooling seemed to serve him in good stead. It seems he could get away with murder (as long as it was committed in ignorance), whereas I was held accountable for the slightest infraction.

❦

THE EVENING BEFORE OUR WEDDING FATHER TRUJILLO *(from Colombia) offered to hear our confessions at Saint Matthew's. Joe was willing but hardly eager.*

I was first in the confessional. I recited all of the ritual prayers and launched into my litany of sins. About fifteen minutes later (I could swear the priest was dozing), I wound it up with a rousing Act of Contrition. My penance was five Our Fathers and five Hail Marys. I approached the altar rail to say my penance as Joe entered the confessional.

About two minutes later Joe was kneeling beside me at the rail. He bowed his head momentarily and then asked: "All set?"

I was midway through my prayers.

"Joe, are you finished already?"

"Yes."

"What did you get?"

(Talk about the seal of the confessional, I even had to know his penance!)

"Five Our Fathers and five Hail Marys."

"How did it go?"

I was about to marry this man. I wanted details.

"Well, when I went in I said, 'Good evening, Father.' He asked me how long it had been since my last confession. I told him fifteen years. He asked me if I was sorry for all the sins I'd committed, and I told him that I was. And he told me to say five Our Fathers and five Hall Marys.

"How'd you finish so quickly?"

"I just said, you know, 'Our Father,' 'Our Father,' 'Our Father,' 'Our Father,' 'Our Father,' then 'Hail Mary,' 'Hail Mary,' 'Hail Mary,' 'Hail Mary,' 'Hail Mary.' "

❧

WHEN IT CAME TO RELIGION, I PROBABLY COULD HAVE USED a little of Joe's nonchalance. The kids could have, too.

Ann would kneel each night in prolonged prayer. Then she'd be out of bed again, four or five times over the next hour, dutifully praying for things or people (pets, even) that she'd overlooked the first time around. With each prayer, she'd do a meticulously precise Sign of the Cross.

❧

I CAN STILL SEE ANN KNEELING NEXT TO HER BED ONE Friday night. She looked extremely concerned for a seven-year-old.

"What's wrong, Ann?"

"Tomorrow's my first confession."

"You'll do fine," I assured her. "You couldn't possibly have done anything worth worrying about."

She nodded, still concerned, and crawled into bed.

It wasn't until years later that Ann informed me of the contents of her first confession. It seems that she wanted to be careful, very careful, not to overlook any of her infractions. So she'd developed a sure-fire plan. She recited each of the Ten Commandments, and then, just to be on the safe side, went on to elicit the number of times she had violated each.

I can just hear her.

"Adultery, six times . . . coveting my neighbor's wife . . . hmmnn . . . three times."

❧

WHEN YOU HAVE FAITH, YOU REALIZE EVERYTHING IS A GIFT. Your fortunes and misfortunes. Your forgotten train tickets and unexpected phone calls. All of it. And probably the greatest gift was our marriage itself.

Gifts are meant to be enjoyed, so enjoy we did. Joe and I would do things on impulse, such as break out a bottle of champagne (we always kept one on ice) at three in the morning on the principle that "quick, we'd better celebrate before something goes wrong."

Or Joe would call me from the office and invite me out to lunch, always at a different restaurant. I remember him taking me to a tiny Korean, family-run restaurant. We were the only customers, with our own dining room and retinue. Joe didn't make it back to the office that afternoon.

Joe had a special tradition. Every anniversary, he would present me with a single red rose and a small piece of jewelry. I suggested once that Joe should send me a dozen roses "like other husbands do." Our next anniversary, a florist's van pulled up at our door. I was presented with a floral arrangement made up of a dozen red roses.

That evening, I heard Joe's car enter the gravel driveway. Standing at the window, I watched him step out of the car, the driver in attendance. He whistled as he opened the door, and I ran to meet him. As he bent to kiss me, he whispered,

"Happy anniversary, dear."

But I'd ruined it. There was no red rose in his hand.

❧

FAITH MADE US ONE. JOE AND I LIVED FOR EACH OTHER. MY day evolved around his phone call, his whistle, his moods.

His moods were contagious. If he was happy there was no stopping me. If he was down I simply joined him at the depths. Any difficult situation we always managed to compound. By the same token, any happy situation became a joyous occasion.

Faith charged our lives, but it exacted limitations. We didn't much mind—a little self-control never hurt anyone.

The Catholic Church is pretty smart.

Take the rhythm method: it can be slow torture. Or it can be an aphrodisiac—abstinence makes the heart grow fonder. Forbidden fruits being the sweetest, all couples committed to "rhythm" are held on a short, but lovely, rein. Once released, they are formidable. I know this much: they seldom tire of one another. Perhaps the success of a happy Catholic marriage lies not so much in its joys as in its joys imposed by discipline.

It was faith that sustained our marriage when times were bad. And times were often bad. But if you believe you're meant to be together, you make it work. No matter what. Not out of duty or even out of love. But out of the surety that God had brought you together and meant for you to remain as one. Otherwise, why bother?

Ours weren't so much the problems of a mixed marriage as those of a mixed-up marriage. Joe was a pessimist. I'm an optimist. I always operate on the principle that "a good attitude attracts good things." Joe was practical. I'm impractical. Joe was factual. I exaggerate. Joe was suspicious. I'm trusting (and usually a pushover). Joe kept his distance. I always rush in. He was discreet. I am too.

From the start, we were never really on sure ground. We never knew where to place the blame when anything went wrong. Friction—and actually really bad fights—were no strangers in our household. Every time Joe would criticize me for being too trusting, too open, too "Irish," I'd eventually retaliate with my weapon of last resort—"and you're a yellow Jap." That would really bring down the house.

Some things that went wrong I actually did blame on Joe's Japaneseness. Japanese men are notoriously mean drunks. They seem to drink for the sole purpose of sinking into oblivion. That was not the case with Joe. But when he did drink, he downed his drinks as if they were Coca-Colas (in three or four gulps), and the effect was immediate, and often bad.

Anything else that went wrong, such as all five of our children having bad eyes (even to almost having the same prescription) we were hard put to know where to lay the blame. Joe and I both had good vision. When I questioned the eye doctor, he explained that eye problems usually skipped a generation. But my mother and father and Joe's mother and Junzo had had no eye problems. That brought us back to the inevitable—his father—and the point of no possible discussion. Joe was fighting his father's and mother's battles all over again, but now I was the culprit.

For Joe, everything that went wrong in our marriage was laid at my doorstep. I was Irish (also Welsh, French Canadian, and Cherokee Indian, but that didn't count). I was Irish, therefore I was impossible.

The rest of our arguments all came down to one thing: his mother. Joe's mother had raised him to be a loner; to be paranoid; discreet. In short, she had bred the perfect spy. Without meaning to, she had determined for Joe a life that spirited him away to the far corners of the globe. The Agency is in her debt.

Unfortunately, those same qualities that made him a really good spy, also made him a really tough-to-live-with husband. He was suspicious of everyone. That's a good quality if you're on the lookout for double agents, not so good if you're supposed to be a loving husband. He trusted no one, and that sometimes included me—the one person he could trust. He always wanted to know where I'd been, but never let on where he'd been. I didn't have a checking account; he doled out money on a need-to-have basis. When it came to friends, especially male friends, he watched my every move. How about the stereo repairman? How long was he at the house yesterday? He even went so far as to accuse me once of having a crush on my obstetrician. (He wasn't far off on that one!)

His mother was still the hot topic of conversation, especially when she visited us in Washington after we got back from Brazil. She made mention of the fact that she had become friendly with a young, Catholic oblate brother from Belgium, who was working at the leprosarium in Molokai. Joe was taken aback. He discouraged the friendship, and said that she should marry someone her own age, preferably Japanese, preferably someone of means—and certainly not a cleric. He asked that if any talk of marriage should arise, she should tell Joe before going forward. She gave him her word.

A short time later, Joe's cousin from Honolulu came by Washington for a visit. One of the first things out of his mouth:

"So Joe, your mother's married, hey!"

Joe looked like he'd been slapped.

His mother had broken her word. She had gone ahead and married the Catholic brother, a man of roughly Joe's age. He'd left the order before their marriage. Joe's inclination was to sever all ties with both of them. He wrote a strong letter stating his sentiments—that from that moment on, as far as he was concerned, his mother was dead.

I objected. What his mother did with her life was her affair.

We could hardly dictate her actions when we were living thousands of miles away. It was a two-way street. She'd deprived us of our property, made an unconventional marriage; but I'd deprived her of her son. Besides, Joe and his mother shared a special bond. He was her only child, her only son. I'd certainly had my problems with Joe's mother, but I didn't want him to sever that tie. He agreed and didn't mail the letter.

Even so, we couldn't even mention her name without it turning into an argument. Once—I think it was when we were in São Paulo—we tried an experiment. I promised not to mention his mother for a period of six months. I mean I couldn't even ask how the weather was in Molokai without it leading to an argument. Joe agreed—no mention of his mother for six months. And that went for him too. I didn't even want to know if he got a letter from her. After all, once you'd read one of her letters you'd read them all. Her health was always bad (arthritis, cataracts) and business at her dress shop was worse—not to mention all the people who had died.

The experiment worked. We had occasional arguments, quickly settled. At the end of six months, I suggested another experiment. Why not go one step further and see if we could mention his mother in a pleasant, offhand way? We couldn't.

There we stood. Joe, glaring; me, hysterical. Both of us selfish. The hell with the kids; it was more important to humiliate each other. Poor kids. I hate to think how hard it was for them. I suspect they simply tuned us out. They must have because they've sure turned out good.

How I wish we could take back some of those painfully beautiful Sundays. Or some of those magical Saturday-night parties, spoiled because we had to plaster smiles on our faces. What was our problem? Couldn't we see that we'd been given so much—great kids, a good career, a life overseas, each other—and were close to destroying it. Couldn't we see that we were tempting God? In our quiet moments, yes. But how do you stop yourself when you're hurt, furious? Count to ten, say a Hail Mary—maybe that helps, but you need more. You need humility.

To calmly await your husband's return, not knowing where he is. To debrief a defector with critical information—those things are difficult. To admit that you're wrong. To ask forgiveness— next to impossible.

But we tried. We tried to fight the good fight. We tried to keep the faith.

❧

CHAPTER XIII

BOMBS BURSTING IN AIR

"HONG KONG?"

"Rome?"

"Don't tell me—PARIS! It's Paris, isn't it?"

The ritual had begun again. Joe loved rituals. It was probably a natural result of his Japanese-Shintoist Catholic upbringing. Carving the turkey, taking family pictures, playing tooth fairy; everything followed a set pattern, with Joe officiating. (I'd just as soon not discuss the turkey, and our many solemn family pictures bear testament to how Joe handled the camera and the family. He did better as tooth fairy.)

But his favorite ritual, next to telling the children his true occupation, was letting them guess what our next overseas post would be.

"Go ahead. No hints."

In this case, a hint wouldn't have been a bad idea.

"Dublin?"

"Brussels?"

"Cairo?"

"Sydney?"

"Manila?"

It went on. And on.

"Calcutta?"

You know you're in trouble when you start hitting Calcutta.

"Is it someplace in Africa?"

The kids were starting to get downright concerned. Joe finally broke the news:

"We're going to El Salvador."

Silence. Quizzical stares.

"Is that in Brazil?"

How Joe ever expected them to come up with El Salvador I'll never know. I'd never heard of the place myself until that afternoon. I'll never forget the expression on Joe's face as David gamely trotted off to get the family atlas and began flipping through it. He finally announced:

"I found it."

And there it was—our new home—tucked away in the elbow of Central America, like some cartographer's afterthought. Even in boldface, El Salvador looked pretty small.

⁂

JOE'S STOCK HAD BEEN PRETTY LOW WHEN WE HIT WASHington in November of 1963. He'd turned down two good jobs and no one was about to do him any favors. Angola had been the first. Joe was to set up a small CIA station there. From Brazil, Joe had sent a one-word cable in reply:

"No."

It struck me that Joe always had a preference for the cushier posts. He'd had a grubby childhood and didn't want to relive it in a bunch of grubby foreign countries. Angola was his mistake. I think he should have taken it.

The second was La Paz, Bolivia, where Joe was to be deputy chief of station. When he refused that, Jim F. was pulled out of Mexico and given the job. He later ended up as Joe's boss.

Always imperious by nature, Joe accepted a nothing job in the

Latin American branch and bided his time, convinced that on-the-scene performance would turn the tide.

Bill B., an old friend and ex-boss from our Japan days, was the head of the Latin American division. He had told Joe to hold out for a chief of station position. But since Joe was still a GS-14, the only countries available to him as chief of station were small ones, like those in Central America. The only one open was El Salvador.

Joe was incredulous. He'd envisioned something far more exciting.

"Besides, Bill, what's happening in El Salvador and, more important, who cares?"

Bill pointed out that it wasn't the country, but the job that Joe should be interested in. As chief of station, Joe would enjoy all the perks and, best of all, be his own boss. After he'd established himself as a chief of station in El Salvador, the next time around he could demand a more important post.

Based on Bill's advice (and what he heard from colleagues about El Salvador's beaches), Joe decided to accept. After Mass one Sunday, he broke the news to a journalist friend who was quick to respond:

"Joe, what in God's name did you do to deserve that?"

❦

WE ARRIVED IN EL SALVADOR IN 1966. THIS WAS THE HEYDAY of cigar-toting Communist Lotharios à la Che Guevara trying to export Cuba's revolution to the rest of Latin America. Che himself was somewhere in Bolivia trying to foment an uprising among the peasants in the hinterlands. Americans weren't too popular—the coup in Guatemala in 1954 was still fresh in people's minds. Everywhere you went in Latin America there seemed to be crowds chanting "Yanqui go home!" Nixon had gone to Venezuela some years earlier, only to have his car pelted with soda bottles and nearly overturned by hostile student demonstrators.

All around El Salvador, things were simmering. Guatemala,

one of Salvador's neighbors, had a guerrilla insurgency on its hands. Two years after our arrival in El Salvador, the U.S. ambassador to Guatemala would be felled by an assassin. Another neighbor, Nicaragua, had been ruled for decades by one family, the Somozas. The Nicaraguan peasants were starting to agitate for land reform. The Hondurans had a major border dispute going with the Salvadorans; Salvadoran farmers had been spilling over into Honduras in unacceptable numbers. Reports of atrocities by Honduran soldiers against the Salvadoran settlers had started to filter in from the border towns.

The basic problem everywhere seemed to be land, and the lack of it. Throughout the region, rich landowners, usually in cahoots with the military, squelched any efforts by peasants to redistribute the land. "Squelch" is one way of putting it.

Americans had had a history of intervening in the region, figuring it was our backyard and up to us to maintain stability. My view was that we'd had our revolution—why not let them have theirs? The usual rationale for U.S. intervention was that Communists were afoot, trying to take control of the government or peasantry. Every beachhead in the region, no matter how small, was important to the U.S. in the fight against Communism. (Remember Grenada?) El Salvador was one small beachhead.

Today everyone has heard of El Salvador since its civil war in the 1980s. But in the 1960s, El Salvador was the Albania of Latin America. Or, to use an example closer to home, Hope, Arkansas, before Clinton came along.

Of course, size isn't what makes a country. Brunei has its sultan, Singapore has its sling. El Salvador, as we discovered, has its *pupusas* (maize tortillas).

We arrived late at night, but the capital city's charms were more than obscured by our reception committee, the man Joe was replacing and his wife. What a pair! They viewed their stint in Salvador as a prison sentence and their hitch was just about up.

Our first ride through the city's darkened, prostitute-lined streets was less than rollicking. With our tour guides, even Shangri-la would have looked depressing. They were quick to point out every trash-encrusted sidewalk, dilapidated building, and impassable road. They seemed to relish it, chortling as they led us down to the next ring of hell.

They made much of showing us Salvador's only noteworthy monument. It was a statue of Christ standing atop the globe, irreverently termed "Christ on the Ball."

By the time we reached the Intercontinental Hotel, the car was pretty much silent, except for the occasional shudder and sigh of one of the children crying. Joe and I weren't far behind.

A "Welcome to Salvador" party was being held that night, and the whole family was invited. Only Mary and David were game. Ann and John and Paul were tired. Joe and I were depressed. We preferred to stay in our room and get quietly smashed. We also had a fight to mark the occasion.

I awakened early the next morning. I looked around—the room was totally unfamiliar. Where had these bizarre Mayan tapestries come from? I hadn't even noticed the room the night before. It was dark, forbidding.

I opened the curtains. Now, there was one view that I'll never forget.

Straight ahead in blazing sunlight was *Izalco*, a massive black cinder volcano that rivals Fuji for sheer spectacle. He looked to be smoldering, irritated that it had taken us so long to notice him. (I was praying it was extinct.) It's visible from virtually every spot in Salvador, but it happened to be situated right outside our window. How fortuitous for the hotelier to have a volcano spring up in the front yard! (In hindsight, I suspect that the volcano had been there long before the hotel.) It was a surreal sight, akin to waking up and seeing an annoyed elephant at your windowsill.

"Joe, you've got to see this! If Salvador has nothing else to offer at least we have this."

Joe, never much for scenery, was impressed. He called room service and ordered a big breakfast for us to eat on the terrace.

I don't recall if the breakfast arrived. I do remember Ann coming to our door and screaming,

"Mary's blind!"

We ran next door and there lay Mary sobbing with pain. She'd neglected to remove her contacts the night before, and both eyes were inflamed.

There's something about children. They have a way of ruining things.

Hurried calls to the Embassy and doctors led us to Dr. Icasza. Mary had temporarily damaged the corneas of both eyes. She sported eye patches and spent the first three days of her stay in El Salvador sightless.

Hotel friendships are much like shipboard romances. They're great at the time but can later become an embarrassment. Salvador was an exception. It was so small that all of us were bound to see a lot of each other. We made friends with the Bastos-Tigres of the Brazilian Embassy. I was especially impressed by Ambassador Pimenta-Buena ("Good Pepper"), Brazil's representative, and his wife. They had lived in the hotel for nine months and were still looking for a suitable residence. Every evening we'd find them ensconced in the hotel lobby watching TV (the only one in the hotel). They'd be dressed in formal attire—he in dinner jacket and she in a long dress and fur stole.

Even the entertainers were a little on the small side in El Salvador. On our third evening, Joe suggested that we stop by the hotel's *boîte* for an after-dinner drink. Kidney shaped, the *boîte* had a baby grand positioned at one end, a mahogany and brass bar at the other, with sparkly lanternlit cabaret tables scattered in between. The floor-to-ceiling windows framed the iridescent aqua blue of the pool outside. It was lovely—actually a little spooky.

As if on cue, the moment we entered, a single chord sounded on the piano as a spotlight came on, illuminating a 5'2", tuxedo-clad Mexican at the mike. His voice was sonorous, unvarnished. His name: Armando Manzanero. With the first notes of "Esta Tarde Vi Llover," the room was his.

Someone once said that music is nostalgia for the future. They must have been listening to Manzanero.

❧

WE MOVED INTO OUR WHITE, BUNKERLIKE HOME WITHOUT incident, unless you count our new driver, René, barreling into a tree on the way. We took a taxi the remaining mile.

Soon thereafter, I was lunching with one of my new Salvadoran friends.

"Oh, Bina," she said, "if they get hungry, they can always just pluck a banana off a tree."

She was explaining the plight of the poor in Salvador, after having just awakened at her usual hour, noon. We were on her patio, framed by thatches of scarlet, fuchsia and orange bougainvillea. Barely out of sight were the smoky, trash-littered, rat-infested, dirt alleyways where the would-be banana-pluckers lived.

She was one of the "Catorce Grande," or fourteen families, that controlled the wealth of Salvador. (When I questioned another Salvadoran friend concerning this, she countered with "actually there are only five." Of course, hers was one of the five.) The fourteen families of Salvador carried their wealth well, but ostentatiously. They outdistanced Brazilians in their conspicuous opulence—everything from Harry Winston jewels to Dior gowns to private Cessnas were in evidence. They'd just as soon fly to Miami to shop, maids in tow, as drive to the family *finca* (country estate) for the weekend.

The whole country was basically their private country club, with the servants put back in their place if they dared step out of line. The effect was of being on a whole other planet—so far removed from anything I had ever known.

The Catorce knew that their world was unreal, but weren't about to let the surrounding poverty spoil their fun. I've never seen such an assortment of uncivic-minded people in my life. Don't get me wrong. These same people were delightful; they were fun; and were rich! I'd be lying if I said we weren't charmed. Great friends—I found the Salvadorans to be the most attractive people in Central America. Once we were embraced as one of their own, their world was ours for the taking: *"Mi finca es su finca."*

I recall complimenting "Lili"—we'd just met—on the beauty of a wicker and mahogany chair in her living room. The chair, obviously very old, had come from her grandfather's farm. The next day, her driver arrived toting the chair and a note:

> *Enjoy it, Bina.*
> *Love, Lili*

These kind of kindnesses made it hard for me to judge them harshly. Besides, they'd never known anything else. They'd never lived on Molokai—or on Harlem Avenue, for that matter.

The views of the "haves" toward the poor may have bothered me, and they certainly didn't square with the Catholic Church's teachings. But for Joe it was far more personal. Here he was, an operative mixing among the powerful and influential Catorce, all the while having at least some understanding of those who were gnashing their teeth waiting to overthrow them. The irony of being embraced by the haves, after having grown up a complete have-not, wasn't lost on him. (Though, in his case, I suppose he could always have plucked a pineapple if he got hungry.)

You wouldn't have known from looking at Joe that he'd never even had shoes until the eighth grade. The cut of his suit and freshly starched edge to his shirt were his cover. (A State Department credential will only get you so far.) Only the superficial think that the superficial doesn't matter. Clothes do matter. They don't make the man, but they make a point. You don't hire an attorney with scuffed shoes—how attentive will he be to your case? And you don't put your life into the hands of an operative wearing a clip-on tie.

Joe wasn't pretentious or overly fastidious in his appearance— no cuff links or suspenders or silk kerchiefs in his breast pocket. Just an occasional collar clip. His look was simple, solid, and un-erringly restrained. His olive-green suit (he had one before Kennedy made it popular) had a slight break at the cuff, so that it draped gently along the tops of his shell cordovan lace-ups. (All of his shoes were from Lloyd & Haig, a tip from a Michigan Law classmate.) His cotton seersucker was a subdued gray, not the poplin, blue-and-white candy-striped variety. (I think he got the idea from my father, who used to wear his seersucker during the summers, with a straw boater.) Joe shunned labels—"Why should I advertise for someone else?"—except for the stray white Lacoste shirt for tennis. (He even went so far as having the dealership remove its insignia when we bought a car.) Or monograms—"I know who I am." The only exception was his black pigskin secretary from Camalier & Buckley, his initials stenciled in a dull gold inside (my anniversary gift to him).

If anything gave Joe away, it was his hands—large, square and slightly battered with clean, blunt-cut nails. Hands that had worked. His palms were a soothsayer's nightmare—a labyrinth of intricate lines breaking here, deeply furrowed there. In repose, his hands had a certain heft and tensile strength. They appeared capable of grappling handily with the unruly (or wringing your neck, for that matter)—hands that were, perhaps, even capable of getting the cap off one of those Tylenol bottles.

Joe was a maverick, but no revolutionary. Yet he understood what inspired revolutions: the frustrations of those self-same banana pluckers. I recall him telling me that while he had grown up poor, he always felt that he'd had a chance. A chance, as an American, to go to a decent school, make use of his talents. Chances that the poor in Salvador had never had—in Salvador if you were born poor, you died poor. It wasn't fair, and it couldn't last. In Joe's view, this was one country hell-bent on revolution.

Being in Salvador back then, I felt much like a guest at a beautiful Southern plantation prior to the Civil War, embraced and entranced by all that Old World hospitality. But in the middle of the waltz, you're able to hear the distant roar of the cannons at Fort Sumter, barely audible above the violins. The music would be forever dimmed—years later, during El Salvador's civil war— with the image of Msgr. Romero crumpled at the foot of the altar, his cassock bloodied by an assassin's bullet.

Joe tried to sound the alarm. In mingling with the Catorce, he'd predict what he saw as inevitable. Maybe it didn't take a genius to predict revolution, just an outsider. Joe believed the only way to avoid an uprising was to begin creating a middle class. Land reform was the logical place to start.

There were a few enlightened souls—three families of the oligarchs, that I know of, who agreed and were trying in earnest to turn the tide; trying to use what they termed "creative capitalism" to give the poor a hand up. Ironically, one of the heirs of these particular few would later become one of the first casualties in El Salvador's civil war. But, for the most part, the Catorce would listen intently, and then, like passengers on some Latin American *Titanic*, cruise oblivious past the squalor on the way to their next party.

And could they throw a party! Gatsby would have been proud. Swarms of solemn waiters in black tie would stand by patiently while you conversed, offering jet-fresh artichokes, mounds of Gorgonzola, slivers of Spanish *jamón serrano*. You had scarcely finished your Salvadoran Pilsener when another, icy cold, would magically appear in a perspiring glass—the mark of any really good party. The smell was always the same: Patou's Joy and Partagas cigars.

They must have already suspected that they were living on borrowed time—it gave the parties the electricity of an unspoken bon voyage. These weren't the everyone-clears-out-at-10:00-to-get-up-for-work parties that Washington has. These were midnight extravaganzas. At the height of the evening, the French doors to the patio would swing open and, on cue, a dozen mariachis positioned outside would strike up "Sabor a Mi." Decked out in their black and silver uniforms and superabundant mustaches, they'd wind their way through the crowd, madly strumming their guitars and, on high notes, jabbing the air with their trumpets. By the end of the night, some polo-playing Argentine ophthalmologist whose name you had already forgotten would invite the family to his *estancia* near Buenos Aires.

Another planet, all right, light-years away, not just from the

banana pluckers but from the outside world, too. When we arrived in Salvador, the Vietnam War was heating up. Student pro-
tests, Abbie Hoffman, the Chicago Seven—a countercultural
revolution was under way a two-hour flight away. Within a year,
Bobby Kennedy and Martin Luther King, Jr. would be dead. Halfway across the globe, the Soviets would thunder into Czechoslo-
vakia under stealth of night. More ominous still, Castro's Cuba
was sending its bereted henchmen out across Latin America to
organize peasant groups—and finding some success. (I'm no friend
of Communism, but if I were a dirt poor farmer, I'd see its
appeal.)

Meanwhile, El Salvador partied on.

❧

JOE HAD STATE DEPARTMENT COVER AGAIN IN EL SALVADOR,
this time as an officer in the Embassy's political section. Joe's
cover as a Foreign Service Officer meant that he had to fulfill two
distinct roles: an embassy role (he was on the Ambassador's
"country team") and that of CIA chief of station. The relationship
is best symbolized by the positioning of Joe's office—it was within
the embassy complex, but set off to the side with its own
entrance.

Joe's duties as CIA chief of station were more important than
his embassy duties. A chief is to the CIA what an ambassador is
to the State Department. The station chief directs the agency's
operations and bases within a given country, or area. His responsi-
bilities are many: he oversees the office staff; controls agents;
writes and edits reports; cables information back to headquarters;
and updates the ambassador on local news, hopefully, before it
happens. It is essential that the station chief get along with the
ambassador and that they operate as a team. If not, he may find
himself, at the ambassador's instigation, declared "persona non
grata" by the host government and shipped home. But the chief's
principal, and in Joe's case, favorite, function is to recruit agents.

In Japan, the venue for agent recruitment had been the firefly-lit dinner party; in Brazil, the late-night *cafezinho* in the *jardim inverno*. In Salvador, it was the quiet conversation over cigars (to keep the mosquitoes away), while standing thigh-deep in swamp water, your 12-gauge Baretta poised for the white-wing doves.

After work on Friday, Joe and John (and David, if he were visiting from the States) would take off on a friend's private plane bound for his *finca*. The hunting would go from dawn till noon the next day. Then lunch—some beef and rice concoction—at the ranch house, with a nap afterward. (Each man snoozed in his own chair out on the patio.) The important discussions took place while hunting. Joe and the contact would distance themselves from the others.

Often their hunting trips struck me as being more like slaughter than sport. Once Joe drove home with a trunk full of freshly downed doves. When he opened the trunk one of the doves hopped out. He was only wounded. I tried to nurse him back to health; I even splinted his leg, but I guess the sight of all of his dead brothers did him in.

The other major recruiting venue was the riding club, where we owned two horses—Ciceron and Cocolino—and boarded them. Salvador marked our only equestrian venture. We couldn't afford it anywhere else.

John was a fledgling polo player and occasional bullfighter. It was at the polo matches that Joe would make his move. All of Salvadoran society would show up. It was like a tailgate party on a college football weekend, with grilled steaks and beer and coffee grown on a nearby plantation. Joe never tried his own hand at polo. There weren't many horses on Molokai.

❧

IT SEEMED THAT OUR FAMILY WAS PRETTY POPULAR, BUT APPARENTLY not everywhere with everyone. In Salvador, we took part in what was becoming another ritual: getting rejected by the local club, the Campestre. (If we accomplished nothing else overseas, at least we fought for the rights of Japanese-Americans everywhere to join country clubs.)

Two months after the turndown, "Larry" invited Joe to lunch. Larry was from Manhattan and had married a prominent Salvadoran and settled in Salvador. He was never Joe's agent but he did provide comic relief such as periodically calling Joe and leaving word that "Lamont Cranston" (the "Shadow") had called.

But this time Larry was serious and asked Joe if he'd heard anything from the club. He was a member of the board and knew Joe had been turned down because he was Japanese. Since the club had not had the decency, or courage, to notify Joe, Larry felt it only fair to tell him. And he suggested that Joe reapply for membership under Larry's sponsorship. Joe declined.

The Campestre's decision regarding Joe would return to haunt them.

⚓

HOW IRONIC. EVERYWHERE YOU WENT IN LATIN AMERICA people were busy decrying racial prejudice in the United States. Especially ironic was the fact that Salvadorans consisted mainly of Spanish and Indian stock—no blacks were allowed in the country. Joe and I were heartened, but a little surprised, when a new officer was assigned to the political section of the Embassy. He was black. We were even more surprised, and disappointed, when he was sent home six months later at the request of the Salvadoran government.

⚓

I CAN'T BE TOO HARD ON EL SALVADOR, THOUGH—IT DID show our kids a great time. John would hunt the white-wings (*ala blanca*) on weekends, play polo after school, take guitar lessons in the evening and cap off the night with a few pages from *A Farewell to Arms*. Friday nights he dedicated to dancing to "Telstar" and "Georgie Girl" at some pretty raucous parties.

Ann was another story. She was becoming interested in boys, so long hours were spent in the study staring out the window and listening over and over again to "Angel of the Morning," to the accompaniment of the silvery splash of the patio fountain outside. Even then Ann seemed bent on dealing with her mystical self. She tells me that she was depressed. I feel bad that I didn't know enough to help her.

Paul started speaking Spanish better than English and began shunning American food in favor of the aforementioned pupusas, fat tortilla-like pancakes filled with cheese, pork or simply toasted and coated with butter. We all turned up our noses at them at first; later on we received a daily order from the local pupusa-maker on the corner, a forty-year-old widow who huddled over a makeshift charcoal fire on the curb outside our house. She kneaded and pounded out delicious roasted pupusas that her three sons would carry in a tin washbasin, shrouded by a white dish-towel, to twelve houses in the neighborhood. I never knew what happened to the pupusa lady and her family after noon.

Paul's birthday parties became a hot ticket item. I'd have a fort built out of palm fronds for the occasion, with treasure hunts extending to the far corners of the house. *Piñatas* also figured prominently. All our friend's children would come as would the society columnists. (You know a country is small when Paul's birthday takes up half a page in the local paper.)

Mary and David, both attending school in the States, would come down for the summers. It was exciting and chaotic, the way happy families always are. They'd get on the Pan Am flight to go home, tanned, grinning and holding on to their tennis rackets. In those days, you still walked across the tarmac to the plane. It was a great time for us.

I can't leave out our pets. We'd always had a lot of animals, but our home in Salvador was a zoo. There was Tony, the Chi-huahua; Peter, the Siamese cat; and two turtles. Gene Joseph, named after a friend of Paul's, was our Christmas turtle—Joe picked him up on the road on his way home from a hunting trip on Christmas Eve. I painted a splashy poinsettia on his shell. Easter week, Joe picked up another turtle on the same road. This one was named Jack. I painted spring flowers on his back. (Everything had to be color-coordinated.) The turtles enjoyed the run of the house and I came to enjoy the sound as they'd clunk down the marble halls at all hours. It was kind of reassuring, like the ticking of a grandfather clock.

Later we added five baby chicks to our menagerie—Easter gifts to the children. Pastel fluffs, they joined the other animals wandering through the house. Next, John won a goat at the American School fair. He was housed in the back garden.

Soon after, John also acquired the notorious coatimundi—a cross between a raccoon and an anteater. It would have been an interesting animal had he not been housed in a tree outside our bedroom window. Despite his intriguing appearance, he was dirty, constantly hungry, and mean. We kept trying to give him away, but he would not cooperate. I rigged up a collar and leash so that he could be taken on daily walks (anything to get him out of the house). Invariably neighborhood children would want to pet him and he would bite them. Finally, tired of observing the two-week quarantine for rabies (and there were three such episodes), I gave him to the vet. He'd seen so much of the animal that they'd become friends.

The charming chickies grew into gangly gray hens. I could've done without them. I said as much to Raymunda, our cook, suggesting that she give them to someone with a chicken farm.

Raymunda wasn't much for suggestions. That night, we ate two of our pet chickens for dinner. The realization dawned on us during dessert, when I noticed that our chicken troop was down to three.

⤙

MEALS ALWAYS FIGURE PROMINENTLY IN RECRUITING. AS A CIA wife, you sometimes feel like a character out of *Guess Who's Coming to Dinner*. Or lunch for that matter.

One Saturday lunch in Salvador stands out. It was the day that Joe's operations in Salvador shifted into high gear, in the person of Gen. José Medrano.

I'd been out shopping, and was surprised when I returned home to see a Salvadoran military jeep parked in our driveway. I found Joe seated in the center patio. He was deep in conversation with a short, squarely built man with rugged features, fair hair, and light blue eyes. He wore combat fatigues with a general's stars on the shoulders and next to his drink, straight scotch, was a Colt .45 with customized walnut grips. Joe introduced me. The general stood up, removing his cigar from his mouth and, with a slight bow, said:

"José Medrano, at your service."

In this simple, yet compelling man, resided the real power of Salvador. He was head of the National Guard, chief of Salvador's intelligence service and literally controlled the president, Fidel Sánchez Hernandez. Medrano had been offered the presidency but had hand-picked Sánchez instead, with the tacit understanding that he, Medrano, would call the shots.

Joe had taken a calculated risk in inviting Medrano over. Known as a loner, he avoided the limelight and had no U.S. contacts. Like General Mourão in Brazil, Medrano was comfortable with Joe. They enjoyed many similarities—humble origins, a certain macho appeal, direct approach and common goals. What was good for Salvador was good for the U.S. and vice versa. Why not combine efforts and work together? In effect, Joe was asking Medrano to be his agent and Medrano accepted. Joe urged Medrano to begin redressing the wrongs of the country by trying to create a middle class and initiating a land-redistribution program. In turn, Medrano asked Joe to provide advice, supply equipment and help perfect the Salvador Intelligence Service. A friendship and working association was born that day that both men treasured—and benefited from.

Medrano was a legend in Salvador. Born poor, the only avenues open to him for advancement were the Church or the military. He was hardly cut out to be a priest so he chose the military, or National Guard. As he rose through the ranks his reputation grew, and with it countless stories concerning him.

His three marriages, mistresses, children, legitimate and otherwise, were fact. As his image emerged—tough and strong-minded people began to take sides. There was no middle ground with Medrano.

He was a showman and never appeared without his .45 on his hip. Theatrics aside, he once confided to Joe that his revolver was a mere prop. He was a lousy marksman and, according to Joe, would "probably have better luck throwing his handgun at a target than shooting at it."

Equal to his fame with a .45 was his fame with a *capucha*. A *capucha* is an instrument of torture that is similar to the hoods worn by Capuchin monks except that it completely covers the head. When drawn over the head and tightened it cuts off almost all air—an extra tug and the prisoner is dead. It was rumored that Medrano once killed a suspected Communist with a *capucha*. Probably untrue—in fact, the rumor probably started with Medrano. But it was pretty effective propaganda. Medrano had only to hold a hood in front of a prisoner to make him talk.

❧

HELPING JOE GAIN ACCESS TO POTENTIAL RECRUITS WASN'T AS much of an issue for me in El Salvador, where the society was so inbred that anyone halfway respectable was welcomed with open arms. In Japan, my focus had been on school committees and cultural activities; in Brazil, on charities and the Church. In Salvador, it was on a mummy.

Salvadorans had a social rather than a cultural bent. Their only cultural institution that I recall was their Anthropological Museum. The building itself was imposing but inside it was a wreck. Nothing was catalogued. The displays were unstructured and confusing. Mixed in with some rare pre-Columbian pots you'd find a moth-eaten stuffed monkey hanging from a tree and a grinning, desiccated macaw perched on a branch. Their most prized possession was the aforementioned mummy. It dated back to pre-Columbian times (that's a guess), and reposed in a glass case at the entrance to the museum.

Marina Sánchez Hernandez, the president's wife, felt that her duty lay in furthering Salvador's cultural heritage. The Anthropological Museum was the logical place to start. She decided to completely redecorate it.

She called on me and four Salvadoran friends to serve on the committee. The committee meetings were a horror, just an excuse for another tea party, and I decided to throw in the towel. That night Joe and I ran into—of all people—Jerry Mumma, an architect who had helped us redesign our house in Washington. He was heading up the U.S. Trade Fair that was being held in the same park as the museum. Talk about a small world. It was like running into your accountant at an Algerian flea market.

I talked him into meeting with one of the committee members, "Chita," and me at the museum to give us some ideas. He was so appalled at what he found that he sat down, right then and there, and drew up complete plans for the inside of the museum, the garden, and even recommended what colors we should use. Under his direction most of the old showcases were junked and new clear plastic cubes were constructed. Having the president's wife as our chairman helped. We had all sorts of carpenters, gardeners and general laborers at our disposal.

Jerry recommended that we paint the outside of the museum an earth brown, the inside was to be white with dark brown felt linings for the cases and white gravel in the garden paths. Twenty fountains were installed in a pool in front of the building. The museum looked great but the displays posed a problem. Salvador boasted no real authority on anthropology. We ladies decided to arrange things as best we could, chronologically. Visual beauty was our criterion.

We had a special plastic case made for the treasure, the sarcophagus. The case occupied the center of the museum, the place of honor. Spotlights played up its importance.

It fell to Chita and me to move the mummy. It was late at night. We were alone in the museum, except for the guards who we later swore to secrecy. It was a pretty creepy task given the circumstances, but Chita and I were game. We simply picked the damn thing up and started to carry it. Midway one or both of us lost our grip and we dropped the poor mummy. It broke into at least ten pieces—ten seemingly unrelated pieces. I rushed home, picked up two tubes of Duco Cement and a bottle of scotch; back to the museum and we were in business. We arranged it in the case as best we could and glued it for good measure. Whether the hip bone was connected to the thigh bone, I'll never know. At least no one seemed to notice.

The Ladies' Committee was later decorated by President Sánchez for its efforts.

∽

MAYBE SALVADOR DIDN'T HAVE THE ADRENALINE RUSH OF São Paulo or the surreal glamour of postwar Tokyo. But it was ours. Our family was still young—there would be time for disappointments later. Maybe we were all living on borrowed time.

We took full advantage. If there wasn't much going on at the office, Joe would call me in the late morning and suggest a date— an afternoon excursion to our favorite place, a black-sand beach about an hour out of town called *Sun Sal*. It became almost our own private resort. We'd come in dripping wet from the surf, pick up a cold *Suprema* at the tiny, thatched-roof snack bar, and then flop down onto one of the hammocks that had been haphazardly strung up from pole to tree.

From there, it was on to one of the rusting Pepsi-Cola tables for the freshest oysters I've ever enjoyed (and, remember, I grew up in Baltimore)—gathered right there in the shoals and shucked in front of us. A huge tray of them raw, accompanied by lime wedges and Tabasco, was our usual lunch, along with an occasional turtle egg.

Time, though, wasn't always on our side—it was on our hands. With less to focus on in the office, Joe turned his attention to me . . . and my shortcomings. So when the parties were over and the kids had boarded that plane, I was the one at fault in Joe's eyes, whether it was money, tardiness, or any of the other usual suspects.

I'd be laughing at the dinner table one day, and fuming alone (and occasionally, crying) in the sitting room the next. I did a lot more of the laughing—we all did—but I definitely remember that sitting room.

It needn't have been that way. The morale in Joe's office was good, and the terrorists not yet on the warpath (as they were in neighboring Guatemala). Happily, his ulcer hadn't flared up. He had every reason to take it easy. In Brazil and Japan, I could always blame his unease on circumstances, be it a disfigured agent of dubious bona fides or a late-night skysnatch in Red China.

In Salvador, there was no ready villain. It was something else. Still with us, like some unwelcome guest, was the specter of who his father was. It may not have always been on his mind, but it always seemed to be in his head.

My response was to try to make things "nice"—to compensate for his lousy childhood, to do what his own mother never could. The kids (and turtles) were always well-turned out, the house, a casbah (to the extent of our means). I was really overdoing it— and getting nothing but more glares from my husband. I'd talk too much at parties, a clear sign of not being sure of yourself. I was nervous, and skinny as all get out. I was worn out.

Who cared about his father? Couldn't he see that it was the Joe he had *become* that mattered—and interested me? Actually, we didn't even talk much about his childhood. Instead, we just talked about the other half of the equation: his mother—which, of course, led to more of my self-imposed solitude in that sitting room.

Here we had attractive, well-behaved children, and he wants to know—still—why they don't look more Japanese. The final straw was the hormone ordeal. The whole thing made no sense. On one hand, Joe was busily force-feeding the kids "tiger's milk"—a distinctly untasty concoction of milk, molasses, raw egg, and brewer's yeast—at the very time he was consulting with a local doctor about the possibility of giving those same kids hor-

mone shots to slow their growth. At first I went along with the
idea, mainly because I was sick of Joe's bemoaning the "Japanese
height" issue and because we were well on our way to having a
family of mini-giants.

I went along with it in theory, but practice was something
else, especially when it came to one of our sons. The vials of
hormones and shot paraphernalia, courtesy of the National Insti-
tutes of Health in Washington, D.C., were carefully stored in our
icebox, all ready for the first weird experiment.

That morning, after Joe left for the Embassy, I threw the
whole business away. Actually, I buried it deep under piles of
pupusas and orange peels in one of our garbage cans. Hormone
time arrived, with Joe ready to do the honors.

I confessed my malfeasance. Joe greeted my fait accompli with
furious equanimity, frightening silence. He was beyond anger.

As we stood there, dangerously still, the only interruption was
Gene Joseph's friendly "clickity-clack" down the marble hall outside our bedroom door. I think it was that "clickity-clack" that
saved our marriage.

Somewhere between that dining room and sitting room, we
both grew up.

⚓

AND SO DID OUR MARRIAGE.

It was on the occasion of our twentieth anniversary that I
wrote Joe's mother to thank her for the gift of her son—I have
to give myself some credit for trying to make things right with
her. I suggested that he might want to visit Molokai to see her
and meet her husband. It had been years since he'd been back.
She evidently agreed that it was a good idea—back came a ticket.
One ticket.

It was old-home week for Joe. He swam in the surf, met with
some of his buddies from the 442nd, most of whom had stayed
in the Islands. He met his mother's new husband, Jeff Swerts.
Joe lazed away the afternoons napping. His mother prepared *lomi-*

lomi salmon and the rock barnacles, *opihi*, that he'd loved as a boy. This was not meant, though, to be a vacation. He'd flown 6,000 miles to ask his mother a question.

Her answer: "Junzo was your father." End of story. No questions, please.

Maybe it was the end of her story, but Joe deserved better. He was entitled to some resolution. For God's sake—for my sake—I wish she'd just talked to him, placated his fears, explained away the nagging questions. You don't get that many chances.

There had been too many rumors for it to end with four words. Was it the Eurasian who had killed himself? Had it been the leper colony surgeon who had vanished one night on a freighter for Shanghai? Who knew?

She did. Joe never would.

༄

I EXPECTED THINGS TO BE PRETTY QUIET WHILE JOE WAS AWAY. At first they were, until late one evening when someone came pounding on the front door calling for the "Madam." When I peered out the dining room window, I spotted National Guardsmen (two jeeps full) on the front terrace, seemingly attempting to surround the house. Fortunately, every window and door on the house was protected with iron grillwork. Raymunda came to my rescue. She was an awesome figure. She matched me in height, but she was almost double my weight. Her size, combined with an ancient wig, booming voice, and vintage .32-caliber revolver made her a fearsome adversary.

Raymunda commanded the guardsmen to halt and informed them I was away, traveling. In the meantime, I telephoned Joe's deputy and asked for help. He called the Marine guards from the embassy. Together they converged on the house. The National Guardsmen thought better of their break-in scheme and left. I was uneasy until Joe returned home. When I questioned him as to the guardsmen's motives, he had a plausible explanation. Perhaps his friendship with Medrano, the head of the National

Guard, had occasioned jealousy. Factions of the government hoped to get at Medrano by throwing a scare into me.

My fears were unfounded—and, my moment of glory short-lived. Careful checking disclosed that our house had previously been occupied by a woman of questionable morals. She'd been just that—a madam.

I knew I had to play a lot of roles as a CIA wife: museum curator, observant hostess, discreet PR agent, inquisitive mother. But I had my limits.

The guardsmen might have tried getting the correct directions from our son, David. David was studying at Gonzaga High School in Washington, and came down for the summers. Of course, he borrowed the car once in a while. The car somehow ended up stalled in a terrible area of town late one night, in front of a notorious bordello. Its name: "Mocambo," a misspelling of the Clark Gable classic. Joe had the embarrassing task of picking up the car the next morning.

Unfortunately, a diplomat's car isn't anonymous overseas; you have CD (*Cuerpo Diplomatico*, or Diplomatic Corps) license plates. Talk about a KICK ME sign.

In Salvador, I'd have traded immunity for anonymity any day. That plate led to more problems than privileges.

❧

BEFORE LEAVING FOR EL SALVADOR, JOE HAD RECEIVED training in defensive/evasive driving techniques. Latin America was rife with terrorism, and CIA operatives had to be on their guard. The training was conducted at the agency's proving ground, "the Farm." Agents were taught how to ram a threatening car (and cripple it in the process) and how to get away in a hurry. Pursuit techniques were demonstrated. Joe enjoyed this phase of training.

Campesinos must be terrorists at heart, because once aroused, they can be just as lethal. We were on our way back from lunch at a *finca* in a remote area of El Salvador. Our host's son had horses stabled at the same riding club as our kids.

We got stuck behind a rickety bus laboriously making its way along the winding country road. Joe passed it but clipped the front end of the bus in the process of avoiding oncoming traffic—the bus had obligingly speeded up once Joe was abreast of it. Evidently, the driver must have noticed our diplomatic tags. The bus immediately repassed us and stopped, blocking the road.

Every drunken *campesino* in the south of Salvador must have been aboard that bus, along with some vocal poultry. The men poured out in unbelievable numbers, all armed with machetes. Joe could see that there was no reasoning with them as they advanced toward our car. No gringo was going to get ahead of them. He told the kids to throw themselves on the floor, secured the windows, and locked the doors. The handgun Joe kept under the front seat was at the ready. By this time we were being surrounded by them.

Joe motioned to the workers in a conciliatory manner, as if he was going to settle with them. Just then, he threw the car into reverse, somehow swinging the car off the road. Then we lurched forward through a ditch and up a hill around the bus. Once on the other side, we got back on the road and headed home, the bus in hot pursuit. It happened so fast that it took me a second to figure out that the bus was now behind us.

Joe applied the evasive driving tactics he'd learned from the agency. The bus driver must have taken the same course—he managed to follow us most of the way home. I've seldom been so scared. I'll never be able to erase the vision of that bus and its crazed passengers barreling down on us as Joe wove the car in and out of deserted, sleepy cobble-stoned streets. Joe finally lost him about five blocks from our house.

✎

MY NEXT ENCOUNTER WITH A BUS IN SALVADOR WAS MORE MY style.

Late in our tour, President Johnson was coming to Salvador

for a three-day meeting with the five Central American presidents. While her husband was involved in a round of diplomatic meetings, Lady Bird Johnson was to tour Salvador in a rented bus. (At least Joe knew who *not* to hire to drive it.)

The day before the Johnsons' arrival, Joe called me from the embassy. He'd just heard about the bus detail and asked me to decorate it, since I had handled the decorations for embassy parties.

Mary, Ann, and I took off for the flower mart in the center of town. They sold every type of artificial flower possible. They were made of crepe-paper, calico, silk, anything you wanted. I wanted ten dozen huge yellow crepe-paper roses. They were specially made within two hours.

I called upon the wives of the CIA personnel to help me decorate. Every one of them turned out. We met that evening in the embassy garage (where the bus had been delivered) and worked until two in the morning. The embassy was really jumping that night. While we decorated in the garage, our husbands worked inside the embassy, and outside, students rioted.

Salvadoran friends had contributed palm fronds from their gardens—some as long as six feet. All told, we collected enough to convert the inside of the bus into a tropical bower—with Texan overtones. It was like going into a grotto made of palm fronds and yellow roses. I wish I'd taken a picture of it.

Meanwhile, Joe was in charge of security measures for the president's visit—no small task given the prevalence of terrorism in the region and considering that six presidents were participating in the meeting. The entire resources of the Salvadoran Intelligence Service and the National Guard were thrown into service. A huge detail of U.S. Secret Servicemen was flown in a week prior to the visit—as well as a large contingent of CIA personnel. Combined, they balanced out the number of guards shipped in for Gen. Anastasio Somoza's (Nicaragua's president) protection. Ambassador and Mrs. Castro turned over the embassy residence for President Johnson's stay.

Preparations attendant to an American president's visit are unbelievable. Everything has to be flown in, from food to drinking water. Special telephone lines are installed. Even the presidential cars for the inevitable motorcade have to be airlifted—all of this mainly to boost the ego of a lame duck president who was being besieged at home because of the Vietnam conflict. This was the last overseas trip that President Johnson made before leaving office.

From the moment they touched down at the airport, Lyndon and Lady Bird Johnson charmed all of Salvador—and the five Central American presidents. Johnson was very much his own man throughout the visit. His was a commanding, compelling presence.

Medrano, as head of the National Guard, was entrusted with the physical safety of the visiting dignitaries. A Secret Service man tells of his concern on leaving the airport when he spotted a jeep with a lone driver riding parallel to the motorcade—and Johnson's limousine. It was Medrano riding "shotgun" for the president. Medrano knew that some unfriendly students were stationed along the way armed with a mixed arsenal of rocks and bags of urine. Spotting them, he raced ahead, drew abreast, and screamed obscenities at them. The students let loose with their whole barrage. Johnson's limousine passed unscathed a minute later. The students had shot their wad.

Lady Bird Johnson is not photogenic; she's better looking in person. Bubbly and charming with her soft Texan accent, she was the perfect foil for her husband. I think she really enjoyed the bus trip—it may not have been Air Force One or the Orient Express, but I think it reminded her of home. As we transported her, she was transported back to Texas.

Their visit was an unqualified social success. The best party was the one given by the Johnsons the night before they left. It was to have been held at the Campestre Club. When Ambassador Castro was consulted, he changed the party's locale to the Intercontinental Hotel, with the statement: "If Campestre saw fit to turn away one of my senior staff members I hardly think it fitting that they entertain our president."

You know it's a great party when Salvadorans leave impressed.
Flowers were brought in from the States, even the flower ar-
rangers. Hope Somoza was the best looking and best dressed of
the presidents' wives. President Johnson shared the honors for the
men—he was also the best dancer. Johnson noticed our daughter
Mary right quick; at six feet, she was stunning. So that evening
my daughter danced with the president.

⚬—

UNFORTUNATELY, JOHNSON'S VISIT DIDN'T DO MUCH TO
improve relations between Salvador and its neighbor, Honduras.
The two countries had never gotten along. Salvador is over-
crowded and some residents had had to settle and farm in Honduras. Otherwise, I can't imagine why they'd want to live there. But
I speak from a limited experience; it was limited to three-quarters
of a day.

We were faced with a long weekend. Joe suggested that we
visit Tegucigalpa, the capital of Honduras—a five-hour drive from
San Salvador. We carried a picnic, as we'd been told that Honduras boasted no roadside stands. What we hadn't been told was
that Honduras (back then) boasted practically nothing—no towns,
houses, nor trees. It was barren. (Having seen Honduras prepared
me for the televised moon landing.)

Our prime concern was finding a tree, any old tree, under
which we could picnic. It was a hot day and the sun was blinding,
the children deliriously hungry and thirsty. David spied a tremendous banyan tree, and Joe pulled off the road. As one, the children scrambled for an iced bottle of Coke. I'd forgotten to pack
an opener. Everybody was furious. Honduras was disappointing,
the ride was too long and dusty (the car had no air-conditioning),
but it was all my fault. The boys were in the process of prying
off the caps against a rock; Joe was uncorking the chilled white
wine, and I was handing out sandwiches and deviled eggs when
Ann screamed. She couldn't speak. All that she could manage to
do was point upward.

We were all faced with a horrible sight. The tree was one giant, writhing iguana. They were draped over the limbs, curled up in crevasses, like so many giant pre-historic snakes.

I don't know who moved faster, the iguanas or the Kiyonagas. We were back in the car and on the road.

Tegucigalpa is perched high in the mountains of Honduras. It's not an attractive town but it enjoys an ideal climate. Clear blue skies combine with a tropic sun and springlike temperatures. We were housed in the Hotel Real (no air-conditioning again).

Our rooms were dusty, buggy and noisy. They fronted the town cathedral whose bells seemed to mark every quarter hour. It seemed a shame to be miserable and have to pay for it.

After a quick tour of the town, we found ourselves back in our rooms. Joe called room service and ordered some cold beer. There was neither. That did it. Joe jumped to his feet, marched to the adjoining room where the children were absorbed in their comic books, and announced:

"Okay, kids, we've done Honduras. Let's head back."

Our next family outing, to Guatemala, worked out better. Joe's friend, Dick Welch, was chief of station in Guatemala. So, of course, we did the neighborly thing—we visited back and forth. We especially liked going to Dick's home on the outskirts of Guatemala City. It was a converted brothel, a distinction I think Dick enjoyed. He had lots of room to put us up, if you didn't mind the red velvet-brocade walls. And not an iguana in sight.

❧

I NEVER THOUGHT HONDURAS WOULD TAKE OUR QUICK EXIT so personally.

I was home alone at about eleven one morning when our personal phone rang (the unlisted one). I recognized the soft-spoken voice on the other end of the line: it was Marina Sánchez Hernandez, the wife of the president and close friend of my mummy conspirator. She made small talk and then casually asked if Joe could get her husband, Fidel (also known as "Tapón" or,

"cork" because of his 5'1" stature), a pair of size-four combat
boots so that he could "go hunting."

Even I picked up on the implications of the request. It had
been rumored that Salvador was close to war with Honduras and
that the countries' troops were beginning to mass at their common border. Since the president was also a general and nominal
commander in chief of the army, chances were good that he
needed the boots to head for the front.

Joe was with the men of the American community in Salvador
(wives were not included) toasting July Fourth with a morning
cup of champagne at the U.S. ambassador's residence. Local dignitaries were invited, and the embassy staff served as hosts.

This information was too good to keep. I jumped into our
blue Chevy and headed for the embassy to alert Joe. As I entered
the imposing residence driveway, I bypassed rows of sleek chauffeur-driven limousines and jockeyed into position at the entrance.
Since I was in my jeans, and a woman, I was noticeably out of
place. I was anxious to fulfill my mission and make a quick, quiet
getaway without getting out of the car. I signaled to Bob, the
military attaché, who formed part of the receiving line that
stretched all the way to the curb. Bob was in his white dress
uniform, clusters of gold braid encircling his shoulder. He surreptitiously summoned Joe who was inside standing near the ambassador. As Joe approached the car, his expression was a blend of
concern and annoyance. He was dressed in his version of a diplomatic uniform—dark pin-striped suit and crisp white shirt that
contrasted sharply with his black wool tie and his dark tan.

He was one handsome man.

I repeated Marina's request in hushed, conspiratorial tones.
Joe seemed more concerned at the prospect of finding size-four
combat boots than he was with the political overtones of the
message. He thanked me and turned to speak to Bob about securing the boots. (Bob cabled Washington, and they were flown in
the next day.) Mission accomplished, I gunned up the car, forget-
ting it was in reverse, and rammed into the Mercedes limousine
parked behind me. I can still hear it: the tinkle of shattering glass,

the crunch of buckling chrome and the amazed gasps of assembled chauffeurs. Joe turned around and now his expression was one of *pure* annoyance.

I hadn't cared to join the party but now the party joined me. Everyone was outside, surveying the damage. I had caved-in the front of the immigration chief's official car. He and Joe exchanged cards in an amicable fashion, and I drove home feeling like a fool.

Joe did foretell the beginning of the so-called Soccer War between El Salvador and Honduras, and even told headquarters the precise time it would start. (It turns out Joe was off by only a few hours.) I'd like to think it was the combat boots analysis I had provided, but it was Medrano who tipped Joe off.

On a Friday, General Medrano marched into the American Embassy. He was a walking arsenal—the ever-present Colt .45 on his hip, a bandolier of shells slung across his chest, and grenades clipped to his belt. He was the head of the army and all systems were go. He asked to see Joe. Understandably startled, the Marine guard passed the buck. He called Joe's office for instructions. Joe proceeded to the lobby and escorted the general to his office where he was advised that the war would break out over the weekend. Medrano was on his way to the front, the Salvadoran-Honduran border, to start the war.

Joe and his office, as well as the whole political and military sections of the embassy, had been following the rising tensions between the two countries. Joe passed on his information to the ambassador and military attaché. They scoffed at Joe's assessment because the official government stance was one of conciliation.

Joe stuck by the report and sent word to Washington to this effect. He had cots set up in the offices and ordered key personnel to dig in for the duration.

Joe's cable occasioned a series of phone calls from the State Department to Ambassador Bowdler, who reiterated his opinion, and the opinion of his whole staff, other than Joe.

They were right. There was no war over the weekend. It broke out five hours off schedule at 5:00 A.M. on Monday, July 12. (Salvadorans are always late.) The first bombs fell before dawn.

Lacking an air force, both governments commandeered private planes and busily pelted each other's country with homemade, hand-thrown bombs. Their very lack of strategy made for a surprising air war. No one was safe—least of all the bombardiers.

Salvadorans were very civil-defense minded. When motorists were told to tape their windshields and dim their lights they were reduced to driving with their heads stuck out the window—the steering wheel gripped in one hand, a flashlight clasped in the other. Lots of accidents resulted.

John volunteered for civil defense in our neighborhood. He was nearly shot by his trigger-happy commander when he failed to give the correct password. He said *"marañón"* instead of *"carañon."* ("Cashew" instead of "stallion.")

I had never been so close to a war—we could actually hear the bombs bursting during the night. The electricity was cut off, so as not to provide the enemy with easy targets. We wandered around the house with candles, like characters out of some horror flick, huddling at my command under the dining room table whenever a bomb exploded. It was kind of fun. Joe was incommunicado, camped out at the office, so I was left to devise a defense strategy. Would the marauding Hondurans overrun the capital? Would looting break out? I surveyed, with dismay, my arsenal: Raymunda and a chihuahua.

The war had started a mere month after Salvador had bested Honduras in a soccer match—hence the Soccer War designation. Some claim the land dispute was actually just an excuse for Honduras to get revenge for the soccer loss. Soccer is taken very seriously in Latin America.

The war lasted about 100 hours. El Salvador won.

⤝

"THAT'S ONE SMALL STEP FOR A MAN, ONE GIANT LEAP for mankind." It was July 20, 1969, as we listened to Neil Armstrong's words as he landed on the moon. We were just about to leave El Salvador.

The setting was perfect. Joe had set up a short-wave radio from the office, with antennas about a yard long, in the center of our starlit patio. Amidst the cicadas and croaking tree frogs, we scanned the warm sky and listened with our family and some invited friends to the historic event.

Leaving a country was as much a ritual as arriving. The last trip to *Sun Sal*, the final dinner at the Mandarin Chinese restaurant, the farewells to Raymunda, René and the others as they stood, in uniform, in front of the house.

We took the dog, Tony, with us; the cat, we gave to General Medrano. I made sure that the remaining chickens actually made it to a chicken farm. Only the turtles were left. We'd become surprisingly fond of them.

Since friends were due to move into our house, Joe and I decided to leave Gene Joseph and Jack for them. Better not to disturb them since they were used to their surroundings.

We packed and moved back into the Intercontinental Hotel a week before our departure.

Five days later, we stopped by the house to say a final good-bye to the staff. We asked about the turtles and were surprised to hear that they hadn't entered the house since our departure. We made the rounds of the gardens in search of them. The poinsettia and spring flowers made them easy marks.

Jack was happily wandering through the strawberry patch. We found Gene Joseph in the far corner under the mango tree—dead. We never told Paul. (I think, at thirty-five now, he can handle it.)

❧

IN THE 1980s, I WOULD READ ARTICLES IN THE NEW *York Times* about Salvador's civil war. It was like reading a personalized obituary column. This friend had been kidnapped, that friend murdered, another missing and presumed dead. Among them was Medrano, gunned down in the early 1980s. Others were friends we had danced with, whose kids our kids had dated.

Familiar names were also reputedly behind the killings of insurgents. That was the unfortunate reality of Latin America. You never knew if the beautifully mannered person across from you might not someday be the leader of a death squad.

❧

CHAPTER XIV

YOU NEVER EVEN
KNEW US

WE LIED ABOUT OUR HUSBAND'S JOBS, STALLED INQUISITIVE policemen, befriended ministers' wives, kept our ears open at parties, deflected the childrens' questions and worried in silence, alone. We were the CIA wives. You never even knew us.

Looking back, I guess one of my chief duties as a CIA wife was that of a one-woman answering service. Once Joe was home, interruptions and phone calls were frequent. Clandestine operations don't observe office hours; the more critical the operation, the more apt it is to happen at night.

It was my job to get to the phone first (we had two lines and four extensions) because it wouldn't do to have the maids know that, say, General Torrijos was calling. The master bedroom phone was on my side of the bed. That way, I could answer, rouse Joe, and give him time to regain his composure before speaking out of a sound sleep. I never screened his calls. I didn't know enough to make judgments. When Joe wasn't there I took the message, got in touch with Joe or a subordinate, so that he could return the call.

I don't care to dispel the James Bond image of the CIA operative. They do have a lot in common—dedication, inventiveness, and frequently, good-looks—but the CIA's favorite instrument is a phone, not a gun.

❧

THE CIA WIVES WHO DESERVE THE MOST CREDIT ARE THE ONES who have to make it on their own, the "deep-cover" wives. Possible covers vary. It can be anything from working for a car-rental agency to running a laundry. The deeper the cover, the harder it is for the wife. Openly, she cannot associate with other Agency wives. She's not automatically accepted by embassy wives nor by the women in whose country she's living—and lack of position can be a real drawback when living abroad. The deep-cover wife is usually resourceful, speaks the language, and involves herself in the civic and charitable affairs of the community.

Sometimes I was drawn, inadvertently, into a friendship with a deep-cover wife without realizing our connection. That's when Joe would step in:

When we lived in Brasília in the 1970s, I hosted the Brasília Book Club meetings each month in Joe's study. In choosing any home overseas, Joe's study was an important feature. Ideally, it had a private entrance so that agents could come and go unnoticed. Our house in Brasília met that qualification.

Joe's study in Brasília was a handsome, wood-paneled room. The furniture was spare, alternately upholstered in rough, taupe hopsacking and sleek black leather. The book club selections graced its shelves—a really eclectic collection of volumes (we even had the *The Joy of Sex*) ranging from archaeology, anthropology, gems, art, history, biography, autobiography, on through fiction.

The club had twenty-four members of many nationalities. There were several ambassadors' wives, businessmen's wives, and three wives from the American embassy. I particularly enjoyed a young Frenchwoman, married to an American, and hoped to make her my friend.

Joe spied her leaving as he arrived for lunch.

"Bina, what's 'Gisette H.' . . . doing here?"

"She's review editor for the book club. Do you know her?"

"I know her husband. Better not be too friendly."

I didn't pursue the friendship.

A deep-cover wife lives with the knowledge that her husband is expendable. If he is caught neither the U.S. government nor the CIA will rise to his defense. He's in deep trouble unless he can talk his way out of it.

When we were stationed in São Paulo, one of Joe's best deep-cover agents represented an American company. He spent half of his day working at his business, at which he was highly successful; the rest of the day, and much of the night, he worked for the CIA. Joe received a call one afternoon from a member of the secret police, another agent, warning him that the businessman/agent was to be picked up by the police for questioning. Joe alerted the businessman, who cleared his house and office of all incriminating files. He was turning them over to Joe at a designated spot in Ibirapuera Park when a cadre of gray Volkswagen police cars pulled up at the agent's house. It was left to the wife to stall for time and feign ignorance and innocence, in Portuguese, until she could contact her husband. He was questioned for six hours. Fortunately, he was Irish, and managed to talk his way out of a potentially dangerous situation.

꒰

THERE ARE WOMEN CIA OPERATIVES WHO I UNDERSTAND have performed with great success. I never, knowingly, had contact with any. I recently ran into one of Joe's colleagues from Japan days. I was amazed to learn that his wife had been an operative. She'd learned Japanese in Korea as a child, where her father was a missionary. This whole time, I'd assumed she was a secretary.

I do know that some wives were actively involved in operations and were paid on a contract basis. I never was. Some took extra risks to insure their husband's safety. One friend, while stationed in Vietnam, disposed of bomb detonators for her husband by throwing them in a river near their home. Another made "drops" for her husband—left messages at a predetermined spot for an agent to pick up.

Unusual services were sometimes required of wives. A colleague of Joe's in Japan once hosted a dinner for some highly placed Communist Party officials. To ensure secrecy, he held the dinner in a "safe house," quarters maintained by the Agency for deep-cover operations, such as meetings with contacts, agents, or defectors. The dinner was staffed by his wife who posed as the cook and actually did the cooking and serving. Another operative, a Nisei, acted as chauffeur, complete with uniform. His Japanese came in handy as he garnered some good information by listening to the guests converse on the way home.

Apart from these "operational" wives, a good CIA wife can claim some credit for her husband's success, but a bad one can ruin her husband's career.

❧

My close friend, "Cathy," had been married to a Navy pilot whom Joe recruited to the Agency in Japan. Along with being an accomplished airman, "Tom" was personally friendly with the Dalai Lama of Tibet and could converse with him in his native Chinese dialect. Because of this, he was assigned to the top-secret mission of getting the holy man out of the Himalayas when the Chinese Communists were threatening him.

Some CIA men fantasize about their idea of what a secret agent should be and get carried away with their job. Tom was one of those. Not only did he neglect to tell his wife how long he'd be gone, he neglected to communicate with her at all in the four weeks he was away. Cathy was on the verge of delivering her fourth child, and she was frightened to bear it alone in a foreign country, Japan.

One afternoon, she came by our house in Kamakura. Her due date was rapidly approaching. She was scared. Alone. She beseeched me to intercede on her behalf with Joe.

I felt for her, but advised against it. This was the hard part; she had to hang in there. One phone call could ruin everything. She begged.

I asked Joe. Could he contact Tom? He was against it, but reluctantly complied. An interfering wife could hurt Tom's career—and, as for Tom, he shouldn't have attempted the mission at such an unpropitious time.

Tom returned posthaste. Cathy was delivered of a lovely baby girl, and that was Tom's last mission. He later left the Agency—and his family.

❧

A CIA WIFE HAS TO BE TRUSTING; EITHER THAT OR GO CRAZY. Her husband's opportunities for infidelities are limitless. He keeps odd hours. He spends much of his time in safe houses. These quarters are a perfect foil for an assignation. A safe house can be anything from an apartment or house to a hotel room. Sometimes it doubles as living quarters for Agency employees; it is staffed, if at all, by servants scrupulously checked out by the CIA. Some safe houses are bugged.

Much of any agent's budget is for entertaining, chiefly in restaurants at lunch or dinner. Who could know whether "M. Fernandez" was a swarthy Latin or a slinky Latina?

I do know of a CIA station chief who lived at home with his wife and children and maintained a second home in a safe house with an Air France flight attendant. He was fired.

❧

IN MY EXPERIENCE, MOST CIA MARRIAGES ENDURED. I KNEW of as many divorces within the State Department and the military as I did within the Agency. The problem of the alcoholic or bored wife didn't plague us much. The Agency subjects each potential recruit to exhaustive psychological screening, which usually weeds out the fainthearted or unstable. I suspect that a background check was done on the wives as well.

There were some CIA wives I thoroughly disliked: the ones who talked too much, and called the office all the time. I feared them too; they were a real danger. They seemed to want to assume some of their husband's importance. Obviously their husbands had confided operational matters to them. The indiscretions of these wives threatened the entire fabric of a station.

I recall giving a luncheon for a group of office wives in Panama, when one of the wives mentioned an apartment that was being bugged and monitored. I was astonished but tried not to let on. I managed to change the subject.

That evening when I questioned Joe, he hit the roof. The bugging discussed that day was highly sensitive. Here he was, chief of station, and he'd never disclosed any of his operations to me. Yet a case officer on his staff, a trusted aide who monitored the bugging, had told his wife.

The next morning Joe called a special meeting of his staff. He cited the indiscretion and warned against future violations. No names were mentioned, but it wasn't necessary. Later Joe spoke with the case officer. He gave him a thorough dressing-down.

I was not entirely immune. I'm naturally talkative and inquisitive. I also like a good story, so for me to play the role of the dutiful CIA wife was difficult. When Joe returned from a mission I would try to question him. He would make that motion—the one that always infuriated me—opening his fingers wide and then clamping them shut. In other words, "keep still."

❧

SOMETIMES EVEN AN EFFORT TO MAKE SMALL TALK COULD GET you in hot water. Joe was traveling once when I attended a black-tie dinner at the embassy of Bangladesh in Brasília. Without Joe there, I figured I had to bone up on the history of Bangladesh, in the hope of impressing my host. The dinner was comprised mainly of ambassadors and their wives and a few Brazilians.

Seating at diplomatic functions, although strictly regimented, can be confusing. The host and hostess can be seated in the middle or at either end of the table. This particular evening, I was seated in the center of a long table. I was flattered since I'd often been "below the salt". (Joe's "cover" in Brazil placed him sixteenth on the U.S. Embassy diplomatic list—low man on the totem pole.)

I assumed that the man across the table was my host, the Ambassador of Bangladesh; at least he looked like him. That was my first mistake. My second was that I insisted on speaking Portuguese (a case of being holier than the Pope since English is spoken in Bangladesh).

I proceeded to quiz the ambassador:

"How many people are there in Bangladesh?"

"Oh, I don't know," he said. "Two, three million."

(Two or three million? Southern Asia wasn't my forte, but seventy-five or eighty million sounded more like it.)

I continued, undaunted.

"When was it that it achieved its independence?"

"From whom?" was his query.

I ventured quietly, "The Indians, I assume."

"I don't know. Was it part of India?"

Now startled, I turned to the lady on my right and muttered:

"No wonder Bangladesh is in trouble. If all their ambassadors are like this they'll never make it."

"He's no ambassador. He's my husband. He'd never heard of Bangladesh until we received tonight's invitation. He runs a textile mill."

Sure enough when the ambassador, who was seated at the head of the table, stood to propose a toast he bore a striking resemblance to my tablemate. In fact he bore a striking resemblance to every other man there.

✎

YOU MAY NOT HAVE KNOWN WHAT YOUR HUSBAND WAS doing, but sometimes you were able to provide some guidance. On a few occasions, Joe consulted me on situations that required moral judgments from him. Once, Joe received orders from Langley that went against his conscience. He indicated as much by return cable. As the man on the scene he felt that he was in a position to question and dispute the orders. His superiors in Washington acceded to his wishes—to the CIA's credit. He never compromised, nor was he forced to.

Joe wasn't always a fan of the Agency's tactics. I recall that he was disturbed to learn that the CIA was surreptitiously opening the mail of some Americans who were corresponding with friends behind the Iron Curtain. His experience with the FBI and his radio in Hawaii had turned him into a civil libertarian in that regard.

❧

THE CIA FAMILY IS UNIQUE. THEY LIVE A LIE BUT, TO paraphrase Martin Luther, they believe in God and country and lie bravely. CIA children are told that their father, or mother, is an embassy official, college professor, Shakespearean actor; whatever his cover demands. Only when they reach a responsible age are the children told the truth. It is left to the discretion of the individual operative when, or if, he tells his children. Many, under deep-cover, never do.

Joe lowered the responsible age as the children, and we, got older and more tired. Joe made quite a ceremony of telling the children. He'd take the child in question into his study, close the door, sit him down and tell him the true nature of his work. Up until that moment, the children, like the rest of the world, believed that their father's "cover" was his true occupation.

Their reactions varied.

Mary was mad, mainly because "Mike Brown" had told her first.

I remember well. We were living in Washington, entertaining Agency people, when Mary burst into the room.

"Daddy, Mike says that you work for the CIA."

Mike's father was a deputy assistant secretary of state for Latin American affairs. Mary was fifteen, and Joe had kept her in the dark too long.

David was thirteen when Joe told him. He was a little crestfallen because he'd been trying to drum up popularity for years by telling everyone that his father was a "secret agent." Now he'd have to think up another ploy.

Ann was worse. She'd never heard of the CIA.

John was delighted. It heightened his father's, and by extension, his own macho image.

Joe lowered the age of disclosure to twelve by the time it was Paul's turn because by then CIA had become a household word. Paul was pleased, but sorry that he couldn't tell "Matt." Matt had been telling everyone in class that his father was CIA; his father was actually commercial attaché at the Embassy.

It was tough on the kids. Have you ever tried to keep good news to yourself? It's the opposite of the Japanese proverb that says that pain is no good without an audience. Here they were, faced with exciting news and they couldn't tell anyone. They were proud of their father's work but couldn't boast about it, or even discuss it. Later on, when the Agency came under attack, they couldn't defend it.

Joe's job was to operate, and he was a born operator. Occasionally, he even viewed the children as junior operatives. Their playmates, teammates, classmates were all fair game—it was through them that Joe could reach the father. If a natural alliance between the children sprang up, so much the better. If not, Joe engineered one.

At the start of every school year, Joe's office would categorize the children's classmates and uncover every scrap of information concerning a parent's job or position, political affiliations, family connections, club memberships, etc.

"David, do you know a boy named 'Carlos O.'?"

"Who?"

"You know Carlos, don't you? He's in your geography class."
(He even knew the schedules.)

"Oh, that fat kid!"

"You can have him over for lunch sometime, if you'd like."

The idea was to meet casually Carlos's parents, perhaps when
they came by to pick him up. (Joe stopped short of asking David
to include the parents in the lunch invitation.) Fortunately, neither David nor Carlos (nor his mother and father) were pressed
into service and our children were never aware of any part that
they played in Joe's recruitments.

Spying is exciting work—it has a ripple effect in the marriage
and in the home. A sense of suppressed excitement prevails. Of
course, living overseas within the CIA framework does require
restrictions on the childrens' lives. Friends dropping by to play
was not encouraged. Slumber parties were out of the question.
(An agent might appear unannounced.) The kids' whereabouts
were monitored and their activities overseen to an extent un-
known to the average family.

Their phone calls were limited. We had to keep the lines open.
Even five minutes could make a big difference. You can imagine
how difficult that was with teenagers.

Our kids were good sports about it. They were not aware
that the constraints imposed upon them were because their father
worked for the CIA. They just thought they had strict parents.

The kids absorbed the best of what each country in which we
lived had to offer because they were determined to. They developed,
like most of their CIA peers, into interesting adults—multilingual,
adaptable, cosmopolitan. For that we have the CIA to thank. The
Agency afforded us the opportunity to travel, and it paid the bills.

The fact that we enjoyed a close family life while we were living
overseas we owe, mainly, to Joe. We never had much money to
spend on the children, so we lavished a lot of time on them. We
always ate dinner with them, at a time when many of my friends
were serving their children first and eating later with their husbands.
Our best conversations took place at the dinner table. It was the

one time of day when we could count on being together. If our children are articulate, I attribute it to those conversations.

It was Joe who bathed the kids when they were small, helped them with their math, told them their bedtime stories. He made them up as he went along. They usually concerned "Sandbag" the Sailor and his sidekick, Chief Humuhumunukunukuapuaha, actually a Hawaiian fish, but the kids delighted in the name. It was Joe who listened to their prayers and put them to bed. I would make the rounds, with Joe, to kiss them goodnight after having finished in the kitchen. Joe and I would take turns later in the evening checking on the kids to make sure they were "still breathing." We were one crazy pair.

Joe and I believed in God, country, and each other; and we tried to pass these values on to our children. They came to admire discipline, diligence, and fun. They knew what it was to be part of a good family. Friends are fine but when you come right down to it, it's family that counts.

The children took pride in their country and knew what it was to be living overseas on the Fourth of July; to watch the colors being raised and "The Star-Spangled Banner" sung on foreign soil. Or to watch the Pan-American Games and smile with satisfaction when the U.S. water-polo team bests the Cubans in a well-fought match. They experienced the extraordinary patriotism of an American posted abroad, even if they couldn't say what their father did. Our family was part of a big conspiracy, but only Joe was in on the secret.

Joe felt that being a CIA station chief had to be the most exciting, rewarding job in the world.

The station chief lives his work and his work is stressful. Agents can defect, governments can topple, heads can roll—all because the chief mishandled a situation, or failed to relay the correct information to someone with the "need to know"—and the President tops that list.

It's a job that leaves a lot of us widows.

✄

CHAPTER XV

HOME AGAIN, HOME AGAIN

WASHINGTON, AFTER EL SALVADOR, SEEMED A LITTLE LIKE A disco on a Sunday morning. Gone were the homemade bombs bursting midair and drunken soldiers at our door. I found myself missing the pupusa lady.

Our house looked surprisingly nice but the neighborhood had changed. Our close friends, the Gotts, of predawn martini fame, had divorced. Our CIA and State Department friends had moved away, most of them to foreign assignments. It just wasn't the same.

More on the club saga: after all those go-rounds in Latin America, Joe had been asked to join Columbia Country Club, one of Washington's nicer clubs.

We still weren't in everyone's club, though. The world order has changed, but back then the Agency was still run by well-born Ivy Leaguers. With some of them, you sometimes got the feeling that if you weren't cut from the same cloth, you never *quite* cut it.

I recall Joe asking some of his Agency colleagues in the neighborhood if he could join their car pool in to work. I could have saved him the trouble. Joe was a little surprised when they declined to include him, citing time constraints. A little offended too.

Another time, an Agency couple invited us for drinks to meet the host's mother. It was for a "drop-by" drink, and so we dropped by. I thought it was a pleasant evening. The next day, when I called the hostess to thank her, she ended the conversation with the admonition:

"Bina, the next time we have you for a drink, be sure and have Joe wear a jacket."

You may not have been part of their world, but you were still expected to play by their rules.

❧

JOE WAS BACK AT LANGLEY, THIS TIME AS BRANCH CHIEF for Central America. He was still involved in the Agency's efforts on behalf of Downey and Fecteau. By this time, Director Helms was pretty frustrated in his attempts to free them. It had now been fourteen years that they had been in that Chinese prison camp, and all appeals to the Chinese for their release had failed. Helms set up a commission to study the problem, and Joe was made a member.

Only Joe, Paul and I lived at home. Mary was attending the University of Maryland (and hating it); David was at Georgetown University; John at the Hill School in Pottstown, Pennsylvania (a school recommended by one of Joe's agents in Salvador); and Ann at Visitation in Georgetown as a boarder.

Four children away at school proved a staggering expense. I started looking for part-time work.

❧

I UNDERSTAND THAT I WAS THE FIRST PERSON EVER TO BE turned down by the Kingsberry Center, a prestigious remedial reading center in northwest Washington, for their training program. Many of my friends worked there. It was natural that I would want to join them.

I was interviewed by Angus MacDonald, the director:

"Your qualifications are in order, Mrs. Kiyonaga. I suggest that . . ."

"Excuse me, Mr. MacDonald," I said. "How long do I have for lunch?"

"Forty-five minutes. Now, if there are no other questions."

He rose. Clearly if there were any more questions they could wait—at least until I was hired.

I decided to press on. There was one other matter to clear up. It had taken me twenty-five minutes to drive to the center, plus fifteen fruitless minutes of looking for a parking space, plus the probable cost of the parking lot where I'd finally left the car. This struck me as being a pretty expensive training program.

"Would it be possible to assign me a parking space?"

Within the week I received my answer from the Kingsberry Center.

My next stop was another remedial reading center called Suburban Education Center. It was run by Nancy G. (ever since my good friend). She took a long look at me, shook my hand and welcomed me aboard—my way of doing business. This time I'd done the necessary research: Suburban Education Center was five minutes from my house and parking was no problem.

❧

JOE GOT A JOB OFFER, TOO—CHIEF OF STATION IN GUATEMALA. We knew the country well from our visits with Dick Welsh. Joe was pleased. Bill B.'s advice about a more exciting post coming along after El Salvador seemed to be panning out.

I wasn't so pleased. Guatemala was dangerous. An insurgent group was terrorizing the country. Kidnappings and assassinations were everyday occurrences. Our lives would have to be highly restricted and closely monitored. I was especially concerned for Paul. He would have to be constantly guarded. No more walking to friends' houses to play. No more riding his bike in the road. Joe tried to placate my fears by telling me that we'd have a bulletproof car.

All CIA chief of station assignments are subject to the approval of the ambassador at post. It's a sensible move since they have to work closely together. It was just Joe's luck to be assigned to Guatemala at the same time that a new ambassador, William Bowdler, was appointed. He'd always liked and respected Bowdler.

One evening Joe arrived home earlier than usual.

"We're not going to Guatemala. Bowdler turned me down."

I couldn't believe it. I remembered the farewell party they gave us the night before we left Salvador—the send-off at the airport. I remembered the parties that I'd helped orchestrate for Mrs. Bowdler. More important, we'd all been friends.

Joe understood. He'd placed the ambassador in an embarrassing position regarding the start of the Soccer War in El Salvador, and the ambassador didn't relish a repeat. There would have been no repeat. Salvador had simply been a case of a new ambassador relying too heavily on his own staff (rather than on CIA) for information and insights. Ambassador Bowdler's real mistake had been in disputing, and going on record as disputing, Joe's most important piece of information regarding the start of the war.

Ambassador Bowdler did Joe a favor.

Another post had to be found for Joe, but they were all filled or spoken for. It took some fancy footwork, and a shift of the chief in Panama to Guatemala, before the problem was solved. Joe was to go to Panama as chief. It was perfect. Negotiations on the Panama Canal Treaty were due to reopen. Panama was the place to be.

But not so fast.

☙

I WOULD SAY THAT THE KIYONAGA FAMILY WERE DOVES WHEN it came to the Vietnam War. (After Korea, and that fiasco, Joe felt we had no business being in Vietnam.) So when David's lottery number for the draft excluded him, Joe was pleased. Otherwise, he'd suggested that David should take off for Canada, since he, Joe, "had done enough fighting for the whole family."

Ann took things one step farther. She rioted on May Day 1971; had her picture taken on the Capitol steps (featured on the front page of the next day's *Washington Post*); and was arrested and jailed in D.C.'s Armory.

I guess the Agency must read the *Washington Post* too. Another oath was soon on its way to Joe. This time, before we could go to Panama, Joe had to pledge, in writing, his support for U.S. policy in Vietnam.

❧

I'D BEEN WORKING ON MY MASTER'S IN EDUCATION, TAKING night courses at the University of Maryland. Our packing coincided with my exams. It was hectic but I passed.

We were on top of the world.

❧

CHAPTER XVI

AT THE CROSSROADS

COLUMBUS DISCOVERED PANAMA IN 1502. BALBOA WENT HIM one better and discovered the Pacific in 1513 by hacking his way across the country through the jungle. Ever since then, every subsequent discoverer of Panama has taken it for all it's worth.

The Spanish conquistadors stripped Panama of its gold and silver. What they overlooked the pirates didn't—except when English pirate Henry Morgan swept into "Old Panama" with fifty men and captured the town. The Panamanians were ready this time. They simply whitewashed their most prized possession—the golden altar—and the pirates left without it. Today the altar can be seen within the relatively humble confines of San Jose Church in downtown Panama. Henry Morgan missed a good bet.

It was in 1971 that Joe and I discovered Panama.

❧

JOE AND I WERE BOTH LATE BLOOMERS. IT WASN'T UNTIL Panama that we really hit our stride. Joe was chief now at an important post—the Panama Canal negotiations were gearing up when we arrived. The Panamanians and our persistent nemesis, the Cubans, were palling around like long-lost cousins.

In El Salvador, Joe had learned the ropes as a new station chief. He was more self-assured now. He knew the pitfalls of dealing with the U.S. ambassador in-country, after being burned by Bowdler. He knew when to take a stand, when to let problems resolve themselves; who and what to steer clear of. He knew enough to trust his instincts.

I was the seasoned Agency wife. I no longer was startled when the phone would ring in the middle of the night. When the head of the Panamanian Intelligence Service showed up unannounced at our apartment at midnight one night, I knew enough not to try small talk and to just get Joe. I even managed to tame my natural inquisitiveness when I saw the presidential helicopter swoop down and pick up Joe on the beach in front of our apartment after lunch one afternoon. No questions asked.

We were down to two children living with us—Ann and Paul—which led to fewer fights over the pinball machine at Christmas time. Between the two of us, the flare-ups were less frequent, the reconciliations less tenuous. We'd said it all too many times before. I was still nervous and sometimes impossible; Joe was still critical and often unfeeling. But we understood each other now. We didn't have to try so hard.

I remember that we were late one night for a James Bond movie at the Balboa Theater. (Joe hated being late for a movie.) A few years earlier, Joe would have made some remark and been out of sorts. One thing would have led to another, and, before you knew it, the evening would've been ruined. But that night, without a word, he took my hand and walked in. (We always held hands at the movies.) I think we'd finally managed to call a truce. Maybe we could finally, as Nana Cady used to say, "take it slow so you can taste the vanilla."

⁌

WE WERE LIVING IN THE HOTEL PANAMA, WAITING FOR OUR condo to be ready. Joe had left for the Embassy and Paul for school, so I decided to go out and do some shopping. As I closed my hotel door, I heard someone whistling at the end of the hall, by the elevator. I love to hear a man whistle, but the best part was he was whistling Joe's favorite song "Try to Remember" (from *The Fantasticks*). The tune carried me along, down the corridor, until I came face-to-face with its source.

He was a tall, rugged Texan, dressed in a buff-colored gabardine cowboy shirt and trousers, complete with a matching Stetson. I smiled and told him that he was whistling my husband's favorite song. He was mighty cordial, removed his hat and introduced himself as being from Galveston. On the way down in the elevator he asked if I'd like to play tennis that afternoon. Why not, I thought. Joe's usually late and it would help pass the time. The date was set for 5:00 at the hotel court.

I showed up looking absolutely adorable in my tennis whites. (I always felt that the best part about tennis and riding was the outfits.) There he stood, looking even more adorable in his. I was struck by the fact that he was a blond mirror image of Joe: 6'4", trim, and some kind of good-looking. Once on the court, the similarity ended. This guy really could play tennis. I mean he was a pro. The game was an absolute disaster, even with ten-year-old Paul's help. I'd brought him along.

Halfway into the match, "Pete" suggested that we pause for a drink. An actual drink. There the three of us were visiting happily with half of the hotel's guests looking on. Pete was in oil and in town on business. There was no ice to break; the conversation just flowed. I couldn't help but think that this was probably the sort of man I should've married. He was open, friendly and didn't once make mention of having a mother—or a wife, for that matter. I declined his invitation for tennis the next evening (by now it was getting dark) and made my good-byes.

Dinner that night consisted of a one-man recitation by Paul on all of Pete's travels, and so on. Joe seemed pretty quiet.

Next day I was enjoying a peaceful morning when, at about 10:30, the door burst open. It was Joe—looking for "Pete!"

≫

SOON AFTER, JOE ANNOUNCED THAT OUR APARTMENT WAS ready ahead of schedule. We moved across town into the lively rabble of Panama City. A lot of Panama had the personality of a bus station with slot machines. No surprise, given that Panama, with its canal, is a real international crossroads and has more banks than anywhere except Switzerland (ideal for laundering that drug money). The corruption was pretty ingrained: everyone's cousin seemed to be on someone else's cousin's take.

Panama is like two countries in one. The Canal Zone is a throwback to the early part of the century: a colonial America of white-washed barracks, mosquito netting (later replaced by high-powered air conditioners) and carefully trimmed mango trees. The top brass lived in Quarry Heights in old wooden manses with ceiling fans and wraparound porches. Mayberry meets the turn-of-the-century British Empire.

Many people view Panama as a half-hour stopover between planes—hot, muggy, and unattractive. But I like to think Panama was a pretty well kept secret—from the coffee and nectarine plantations of Boquete to the ruins of Old Panama; from the square-trunked trees to chirping golden frogs. It wasn't our favorite country overall, but I liked the people a lot. And at least they had a middle class.

Japan had intrigued us. Brazil had embraced us. Panama— well, it grew on us. Kind of like its native music, the *tamborito*. (Joe termed it a "jig"—"when you've heard one, you've heard them all.") By the end of our tour, we'd gotten to actually like the *tamborito*. It is more subtle than the samba and less exciting than the tango, but it has merit. It is distinctly Panamanian and, God knows, they've produced little else.

In El Salvador, the ruling class were the Catorce; in Panama, they were nicknamed the *rabiblancos* ("white tails") because of their white, European roots. But the real power in Panama, while we were there, resided in the person of Gen. Omar Torrijos, the head of the *Guardia Nacional*, or National Guard. And in another man, the head of the Panamanian Intelligence Service, or DENI. More about him later.

✄

BY GRACE OF GEOGRAPHIC LOCATION, PANAMA IS AN IDEAL spot for a canal. It's not a fact that's lost on the Panamanians. One modest sobriquet: "Panama—Heart of the Universe." Another: "The Land Divided, the World United." My favorite: "A man, a plan, a canal, Panama." (Try spelling it backwards.)

The French got the idea first. They came along in 1880 to try to build a canal, but failed. (The mosquitoes got to them.) Then we Americans gave it a try, after getting a "concession" from Panama for the land. (As someone once put it: "We stole the land, fair and square.") The canal went into operation in 1914.

Americans are pioneers. It's a heritage we show in our Brooklyn Bridges, space shuttles and information superhighways. And in the Panama Canal, an engineering marvel. It's been termed one of the unlisted "Wonders of the World," and I don't question that. Through a contraption of locks, or compartments, a ship is actually raised several levels so that it can reach one of the world's largest man-made lakes and then wind its way across the isthmus, only to be lowered to sea level by more locks at the other end. Simple, ingenious, and mammoth in scale.

Vessels of every conceivable variety and size traverse the canal. While we were in Panama, one massive Japanese tanker barely squeezed through the locks, with only inches to spare. Small ships would pass through the canal in tandem—for as little as $24.00. (The toll amount was based on the amount of water the vessel displaced.) A distance of fifty miles, spanning a continent and connecting two oceans, for the price of a box of cigars. The lowest toll on record was the thirty-six cents someone paid in 1926 when he swam through the canal.

๛

JOE HAD A LARGE, OFFICIAL CIA CONTINGENT UNDER HIS command in Panama, as well as others who were deep-cover employees. His cover was "special assistant to the ambassador;" in fact, he had more people under him than the ambassador. He had two offices: one in the embassy, and one in the Canal Zone. Most of the CIA employees were housed in the latter office. (It was the only building in the area without any windows.) Joe divided his time between the two locales.

It was while we were in Panama that the negotiations for the return of the canal started. Given the strategic importance of the canal, and the American military bases there, the fate of the canal negotiations was vital to U.S. interests in the entire region. Joe's main focus, as chief of station, was to monitor the status of the negotiations. Suffice it to say that the CIA—and, in turn, headquarters and the White House—were always current on what was happening behind the scenes in the negotiations.

Joe was in a tough spot. He felt that the canal, for all of its grandeur, was an anachronism. It was time that it belonged to Panama, but he didn't like the carnival atmosphere in which the negotiations were being carried out. The talks were held on Contadora, a resort island off the coast of Panama (and later, the exile for the Shah of Iran). Conciliation was the theme. The American negotiators Ellsworth Bunker and Sol Linowitz were seen sporting T-shirts emblazoned with the motto EL CANAL ES NUESTRO (The Canal is Ours) and with the Panamanian flag displayed as part of the logo. The inference was clear, given the use of the Spanish language and the Panamanian flag. I find it hard to picture Ambassador Bunker, who was a courtly figure, being a party to such nonsense, but my source is reliable.

The Panamanian negotiators were setting the tone, and the U.S. negotiators were following their lead. Rather than dealing from a position of strength, the United States was eagerly knuckling under. Joe summed it up one night at a party: "They're kicking us in the balls—and we're loving it." Fearing adverse world opinion, the U.S. was intent on hammering out a treaty, no matter what the price.

Take Moisés Torrijos, the brother of General Torrijos. Moisés had had some run-ins with United States drug enforcement authorities. The CIA, trying to put an end to his drug dealing, set up a bust. Instead of cooperating, the State Department evidently intervened—and tipped off General Torrijos. Once alerted, the general had his brother return home from Spain (where he was Panama's ambassador) by way of the Caribbean, rather than through New York. Moisés avoided prosecution, and the matter was hushed up. Nothing was to upset the treaty negotiation "applecart."

I thought Joe's personal view toward the whole Canal issue made more sense. "Why don't we sell the canal to Panama? We built the thing. Why pay them to take it off our hands?" He thought the Panamanians saw the Canal Zone mainly as valuable real estate, instead of a unique, man-made achievement. He predicted casinos dotting the shores of the canal if Panama took it over.

I think Joe was just getting sick of the whole anti-American attitude. Our apartment was only minutes from a notorious slum known as *Salsipuede* (translated, it means "Get out if you can."), the childhood stomping grounds of the boxer Roberto Duran. While we were in Panama, Duran had the habit of decking—I mean, really creaming—any American challenger. The whole country would explode in celebration. Their man had beaten the gringo once again! Cars would drive by until the wee hours with drunken revelers shrieking PANAMA!! PANAMA!! It was kind of irritating—that, and the YANQUI GO HOME sign spray-painted on a wall across from the Zone.

You can't really blame the Panamanians, though. The canal is the only thing the place has going for it. (Oh, I'm sorry, I forgot. Panama also had a cement factory.) The contrast between the picture-perfect Canal Zone and the chockablock chaos of Panama, divided from one another by a tall fence, only added to their sense of injustice.

❧

THE VIEW FROM OUR APARTMENT WOULD'VE TRULY IN-
furiated some Panamanians. Perfectly poised on Paitilla Point,
overlooking the Bay of Panama, our apartment was the wonder
of our world. From our terrace, we could see the ships on the
horizon each day lined up to enter the canal. Rusting Honduran
tuna vessels, massive Japanese oil tankers, gun-metal gray U.S.
battleships—country floats in our own private parade. We once
counted twenty-four ships. Jacques Cousteau's *Calypso* was
berthed beneath our terrace for five days, although I never got
around to seeing it up close or meeting its captain. (A trait that
runs in the family—John spent a year in Barcelona and never
managed to make it to the city's famous Picasso museum.)

At night, the only evidence of the ships was the blinking red
or green lights on their bows, like a distant strand of shimmering
Christmas tree lights. The constant noise and bustle of the nearby
traffic made our quiet perch with its delicate breeze all the
more lovely.

Sometimes, we'd spot Jack, our chocolate brown poodle, am-
bling alone along Avenida Balboa—Panama's magnificent prome-
nade that fronted the Pacific. Jack was "cock of the walk" until
Joe would whistle for him from our terrace (a whistle that needed
to carry half a mile). Jack would stop, perk up his ears, turn right
around and head for home. Within a few minutes, the doorman
downstairs would buzz up to say that Jack was on his way up in
solitary splendor in the elevator.

The sea itself was the best show—we kept an Asahi-Pentax
set of high-powered binoculars at the ready on the wet bar. The
salt-scented air swirled about us, alive with swooping frigate birds,
diving pelicans, soaring coal-black, hideous turkey buzzards and
Paul's paper airplanes. At high tide, the waves would slam into
the volcano-rock caves onshore, and then resurrect themselves a
moment later in a spectacular geyser. We'd spy a leaping porpoise
here, an elusive shark fin there. It felt as if we were back on the
deck of the *Cleveland*—that same blazing heat and gentle stillness.

We lived on that terrace.

On Saturdays it was where we enjoyed our *sancocho*. The national dish of Panama, *sancocho* is a type of chicken stew, consisting of tuber-root vegetables—*yucca*, *yame*, *otoe* (I have no idea what the latter two are) and, of course, the chicken. Tradition has it that in order to insure a first-rate *sancocho*, the chicken should be stolen from a neighbor. We served it on the terrace with bullshots, a drink made with beef bouillon, pepper, lime, vodka and a dash of Jamaican hot sauce. It was strictly a family affair.

≫━

BESIDES THE CANAL NEGOTIATIONS, JOE'S OTHER FOCUS WAS on the Panama-Cuba connection. General Torrijos made many trips to Cuba, at Castro's invitation. Joe was kept informed; an agent always went along. Joe not only got information, he usually also got a box of Cuban cigars with GENERAL OMAR TORRIJOS in gold lettering on the label of each cigar.

The general was a valued contact of Joe's. Torrijos liked Joe. He found the other Embassy officials *débil*, or "weak." Joe found Torrijos to be crude, shrewd and fairly likable.

Torrijos was a Medrano-like figure, a military strongman of humble origins—definitely not a *rabiblanco*. His presence was everywhere. Billboards with his picture graced the downtown area with slogans like: PANAMA: STANDING OR DEAD—BUT NEVER ON ITS KNEES. Torrijos was impulsive. When he wanted to discuss something, nothing stood in his way.

I recall one lunchtime when Joe seemed to be in a hurry. He passed up his usual twenty-minute siesta and left for "the office." A few minutes later there was an awful racket just outside our balcony. I was surprised to see Torrijos's white presidential helicopter land on the sand. I was even more astonished to see Joe hop aboard just before the helicopter took off. I didn't worry. (I'd come a long way from Joe's first unexplained disappearances in Japan.) I knew exactly what to do: wait, and keep my hand away from that phone.

Four hours later Joe was back—by helicopter.

Another time, Torrijos sent his helicopter for Joe while we were vacationing in Taboga, an island fifty miles offshore. Once overhead the pilot thought better of it. The beach was crowded with bathers and it would have been difficult to land. More important, it was far too public.

❧

WE HAD NOT BEEN IN PANAMA LONG WHEN NIXON MADE HIS historic trip to Peking in 1972 to open relations with China.

The commission on which Joe served had been trying to secure Downey and Fecteau's release. They'd eventually reached the conclusion that nothing short of presidential intervention would work. This was their chance. Along with Nixon went the hopes for Downey and Fecteau's release. Nixon did raise the issue in his discussions with Mao.

It was shortly after Nixon's trip that I found Joe on the terrace one evening with a scotch, no lights on.

Fecteau had been released.

❧

ASIDE FROM TORRIJOS, ONE OF JOE'S MAIN CONTACTS WAS THE head of Panama's secret police, the DENI. This man would later attain international notoriety as a reputed drug dealer, murderer and dictator. His relationship with CIA would become the source of much speculation. The United States would eventually invade Panama, taking the lives and homes of many innocent Panamanians in the process, just to capture him.

I remember the first time I met Manuel Noriega. It was after midnight. There was a persistent knocking at our door. Lucy, our maid, heard it first. When she opened the door, there he stood. She recognized him from the papers.

Lucy, usually placid if not downright sluggish, was at my door with an urgent, excited whisper: "Dona Bina! Dona Bina!"

I put a robe over my nightie and went out to investigate. There he stood in mufti.

I knew enough not to ask questions. I ushered him in and went to rouse Joe. He and Joe went out on the terrace to talk. A lot must have happened out on that terrace.

I don't know if Noriega was an agent. I read newspaper reports, during Noriega's trial here in the U.S., that Noriega was put on the CIA payroll starting in about 1971, which coincides with our arrival in Panama. His first contacts with CIA apparently preceded that. Was Joe the one who put him on the payroll? Joe never let on to me in his recitations from his hospital bed. Perhaps he was protecting his "source." All I know is that Noriega showed up at our apartment door unannounced more than once. And that he helped my son.

※

JOE AND I WERE ON A SHORT TRIP TO NEW ORLEANS AND HAD left Paul at home in the care of Joe's secretary. Paul would frequent the beach in front of our apartment looking for shells and whale ambergris (a futile habit he'd picked up from his dad). On one occasion, a man approached him on the deserted beach, exposed himself and then caught Paul and threw him to the sand. Paul fought and managed to get away unmolested and unharmed, minus his wristwatch. He didn't tell Joe's secretary, and mentioned it only casually when we returned.

The *Guardia Nacional* were summoned. Joe and one of his CIA staff, both heavily armed, spent several evenings combing the beach with Paul in search of the man. He was gone, apparently. The Guardia detective suggested Paul keep a periodic lookout over the beach with our high-powered binoculars.

Several months later, when he was home alone from school, Paul spotted the man on the beach. He telephoned for the Guardia, who arrived in a paddy wagon to find a nine-year-old in a Lacoste shirt, arms akimbo, waiting for them in front of our apartment. They went, after some urging, to the beach. (I don't think the Guardia knew what to make of some American kid insisting that they immediately go and arrest some beachcomber.) They were even more leery of Paul when they caught up with the man, who laughingly played dumb.

But Paul had remembered that the man had a large S-shaped scar over his left knee. The Guardia had him pull down his jeans. There was the scar.

Two days later, Noriega summoned Paul to his office for a meeting. (Think of a Latin J. Edgar Hoover interrogating Macaulay Culkin.) We waited outside. Noriega asked Paul some questions: "Was this the man who attacked you? How do you know? Are you positive?"

Apparently, Noriega was satisfied by Paul's powers of observation. He informed us that the man, who had a lengthy criminal history, was being sent to Coiba for four years, a prison island surrounded by shark-infested seas.

≫

PANAMA BOASTS OF BEING THE CROSSROADS OF THE WORLD. For the CIA, at least, that was true. We had many visitors from headquarters during our stay in Panama; usually, it was a stopover for CIA personnel en route to other destinations in Latin America. (Funny how the colder the weather got in the States, the more visitors we seemed to have.)

The traffic was especially heavy right before one major event in Latin American history, an event that would come to mar CIA's reputation, incur the wrath of Congress and lead to public outcries for more scrutiny of the Agency. The event was the overthrow of the Salvador Allende government in Chile. It's been alleged that CIA engineered the coup. Joe never commented to me either way.

The visitors weren't always just on stopovers. I recall a group of five visitors who came to Panama on TDY (temporary duty).

Their visit coincided with one made by Bill N., the DDO—deputy director for operations. Bill was our house guest and we planned a dinner for him. All of the office, and spouses, were invited, as well as the TDY visitors. Despite the fact that Joe couldn't talk shop with me, whenever something was about to break I could sense it. I sensed it that night.

I always gauged the success of a party in terms of intake and output. There seemed to be a direct corollary between the amount that guests ate and drank and the conversation that ensued. Interesting guests helped, and their numbers made a difference. By that criterion ours, that night, was a huge success.

Sometime during the evening the party shrank to about half its size, without anyone even saying good-bye. I thought it strange. Men left without their wives. Joe's chief of operations, his deputy and the five TDY visitors were among them.

Joe took Bill to the airport. While he was gone he received a phone call. The caller's message was terse:

"Tell Joe we're all set."

Joe called from the airport to ask if there'd been any calls. He didn't try to hide his satisfaction and relief when he heard the message. The visitors were technicians sent from Washington, experts in the art of electronic surveillance. They'd managed to install highly sensitive recording devices in an especially critical location.

❧

OUR FLOW OF GUESTS FROM HEADQUARTERS CONTINUED. I think the polygraph operators came next. Their mission: to test the truthfulness of a Communist defector.

Every operative's dream is to snag a defector. I know of one who received a double promotion on the strength of it. Friends of ours were stationed in Africa when the unbelievable happened. A Russian diplomat, whom they'd met at a party, simply marched up to their front door and announced to the wife, when she answered his knock, that he wanted to defect. She invited him in and, during the two hours that it took her to locate her husband, their nine-year-old son engaged the diplomat in a game of chess. The Russian won.

The only instance that I know of when Joe was involved with a defector was in Panama. (There were probably others of which I'm not aware.) This man was an Eastern European visiting Panama as part of a military mission. He approached a colleague of Joe's and requested asylum in exchange for information. Joe was informed and, as chief of station, contacted headquarters, which, in turn, dispatched two polygraph experts to probe the defector's bona fides. By the time the polygraph men had tested, retested, analyzed, reanalyzed, and disputed each other's findings, the Eastern European colonel had lost interest and rejoined his own group.

The final contingent to visit was a team of specialists in bugging devices. CIA sent a team down every six months. Their job was to "sweep" our house, those of other office personnel, and the office, to check for possible taping devices. The phones were also monitored. The team usually consisted of two men plus surprisingly compact electronic equipment. It was up to me to see that the servants were sent off on errands and I would absent myself and the children as well. The sweep lasted about an hour. No bugs were ever found at home. I don't know about the office. I've sometimes wondered if our cars were checked.

❧

BUT WHEN I THINK BACK ON PANAMA, IT'S NOT THE DE-fectors, or prison islands, or helicopter pickups that I recall. I mostly recall *molas*.

Molas are the pride of the Kuna tribe, one of three tribes in Panama. The Kunas originated and live on the San Blas Islands that dot the Atlantic Coast. They are the most peaceable, colorful, and commercial of the three tribes. They do a brisk business with their *molas*, a type of brightly colored, reverse appliqué needlework.

I became addicted to the *mola*. Compulsion is too mild a term. The Kunas would make forays into town and set up shop on the street that separates the Republic from the Canal Zone. It used to be called Avenue Fourth of July; it's now called Avenida de los Mártires (Avenue of the Martyrs), in reference to the looters who died there in the 1964 riots. Joe swears that one day I approached an old matriarch (about ninety), who turned to her daughter and hissed:

"*¡Oye, es la gringa pelirroja!*" ("Hey, it's the redheaded American.")

One particular *mola* stands out in my mind. It was intricately worked, beautifully shaded and looked like a couple of mountains. I wasn't quite sure though. I asked a fellow shopper if he could figure out what it was supposed to represent.

"You can't see it?"

"No," I answered.

"I believe it's supposed to be a brassiere."

He was a distinguished-looking man. He barely flinched. I did. There it was for all to see—a bra—down to the hooks and eyes. Those Kunas were a match for the Japanese when it came to the art of copying. I bought the *mola*. It makes a handsome and comfortable pillow.

❧

I LEARNED ABOUT THE KUNAS BY HELPING PAUL WITH HIS fourth-grade social studies homework. Which got me thinking: maybe I should go back to trying my hand at teaching.

Undeterred by my experience with the Kingsberry Center in Washington, I began substitute teaching at Balboa High. I was called on one morning at 6:30 A.M. (the worst part of the job) to teach English literature. They must have been really desperate. One of my pupils—Ann! I put on the record version of Olivier's *Hamlet* (that'll keep them busy), and then asked Ann to monitor the class while I slipped out to catch Paul's debut in *The Prince and the Pauper.* I'm not sure what I was thinking. The room resembled an Attica prison riot when I returned.

At least I was up front about it with the students in the carpentry class:

"Kids," I said, "I know nothing about carpentry, and these machines terrify me."

I was talking about ceiling-high electrical drills and saws.

The boys were quiet, frighteningly quiet.

"Let's make a deal. You take care of the carpentry, while I watch the door."

They agreed. For two weeks, I caught up on my reading while the kids built all sorts of wonderful things. I guess.

I fared better in my second job: helping out at Mansión Floral, the local flower shop. (Those ikebana classes in Japan came in handy.) It was there that I made a friend.

Her name was Gloria—Gloria Altamirano Duque de Mendez. And she was glorious. My best friend in Panama, and one of my best ever. Loyal, fun, tough and bustling—she swept you up. Even an offhand comment by her was an adventure: "Bina, tell John to beware in the jungle. There are many wild porks." (I believe she meant wild pigs.)

In part, she made me realize how unequipped I was to deal independently with the practical world. She was financing a business; I'd never even written a check. She was arranging flower shipments to Miami; I was trying to figure out how to get a substitute teacher to substitute for me.

After we left Panama, Gloria came up to Washington often with her lovely daughter, Gabby. Gloria died in 1997. I miss you, Gloria. Life just isn't as much fun without you.

❧

I RECALL RETURNING FROM MANSIÓN FLORAL ONE EVENING TO hear Joe announce that we had to leave Panama within twenty-four hours. He'd been declared persona non grata by the U.S. Ambassador, Robert Sayre.

Joe had always enjoyed good rapport with Ambassador Sayre. They respected each other as professionals. Joe made it a point to consult with him on any sensitive material before notifying headquarters. But while Joe was in Washington on consultation, his deputy saw fit to bypass Sayre. Some critical intelligence had been passed to the deputy and he notified headquarters directly— I suspect that he was out to garner some glory. The repercussions caught Sayre in an embarrassing position and he was, rightfully, incensed. Joe had had nothing to do with the incident, but he was responsible for the actions of his staff.

Joe cabled headquarters outlining the case and asking for instructions. A return cable stated:

> JOE,
> DON'T WORRY.
> DICK

Director Helms's interest paid off, and we stayed in Panama. CIA has a way of backing up their people in the field.

Joe was appreciative of Director Helms's support, maybe even a little surprised. I remember that once, after Joe had come back from an earlier trip to headquarters, he commented to me:

"I like Dick Helms just fine. I just wish that he'd quit calling me George."

❧

ABOUT ONE YEAR AFTER DICK FECTEAU'S RELEASE, JACK Downey headed home. Apparently, Jack was released in Red China and walked across the bridge into Hong Kong—and to his freedom. That was his Long March.

Once back in the States, the powers that be at the Agency offered Jack a very lucrative position to reward him and to keep him on.

Jack's reply:

"Thank you, but no. I don't think I'm cut out for this line of work."

Jack, a Yale graduate, applied to two law schools—Yale and Harvard. Yale turned him down, but Harvard accepted him.

He had spent twenty-one years in a Chinese prison.

❧

ONE OF THE MAIN ATTRACTIONS OF LIVING OVERSEAS IS THE adventures your kids can enjoy, whether it's cruising through the Panama Canal on a nuclear submarine or eating a monkey with a Choco Indian chief.

The Chocos live in the Darién Province of Panama, the jungle area. With the exception of their faded CERVEZA PANAMÁ T-shirts, they look and live like something straight out of the Stone Age.

Paul was ten when he and "Christopher S.," a family acquaintance from Washington, spent three days with a Choco chief—and his three wives and seven children. The accommodations were pretty basic: a thatched roof covering an elevated wood floor. No walls. A notched log was used as a ladder.

The chief must have been on some miracle diet. He was reputed to be at least sixty, but Paul said he appeared to be thirty. The chief wasn't sure. The Chocos have no written records—not a problem since they don't read, either.

His proudest possession was a transistor radio, given to him by a United States military man who found his way into the Choco camp while wandering through the jungle on a "survival course." The chief carried it with him everywhere; the radio played constantly, day and night. Paul finally figured out that the chief didn't know how to turn it off.

Their meats ran to iguana, which Paul enjoyed—"much like chicken." On the final evening, the chief, as a special honor, shot a howler monkey with his blowgun and invited Chistopher and Paul to feast. Given the monkey's status on the evolutionary chain, it was a little like eating your third cousin.

❧

DESPITE THE SOMEWHAT UNUSUAL SETTING IN PANAMA, IN many ways we were just a typical American family. Ann was skipping classes at Balboa High School to go to Coronado, the Palm Beach of Panama. John was at the Hill School, regaling his friends with stories of Salvadoran polo matches. David was at Tulane Law, enjoying New Orleans oysters with his future wife, Dede, at Tujague's.

Typical family, maybe, although I don't know how many husbands employ the resources of the world's most sophisticated spy organization to check up on their wives' old beaus.

Joe had just walked in one evening from work:

"Bina—George Cauthen died today."

George Cauthen? It took a minute for his name to register. I rewound the tape twenty-five years and there he was. Actually, there was his car: that black Packard convertible, with a rumble seat, that he'd drive around Bogotá. I tell you, George was fine, but that car was irresistible. I'd step up on the running board, and climb into the front seat and enjoy the inviting cocoon of its glove-soft cream leather interior. I'd accompany him to the airport sometimes to see him off on his flight to Miami. As he'd walk across the tarmac, he'd casually toss the car keys to me over the heads of the crowd.

I can't imagine why Joe, and the Agency, kept track of George all those years. What business did he have tracking George? Sure, George and I had been "engaged," but that hardly merited anyone putting man-hours in on his happenings.

I was saddened to hear that George had been married and divorced three times; had one child; and had died an alcoholic. I was angry with Joe, despite, I'll admit, being a little flattered.

❧

IT WASN'T JUST ME. THE AGENCY WAS ALSO PART OF THE KIDS' lives, in ways they never suspected.

The Boy Scout Pine Wood Derby was a father/son event. The whole idea was for Joe and Paul, together, to build and race a small wooden car. Most father/son pairs labored for weeks, debating the grade of graphite for the wheels, the most streamlined design. Paul had inherited much from his father, including Joe's ineptitude with handicrafts. Neither of them was remotely interested in the derby. Joe ended up enlisting a master craftsman on CIA's payroll to help with the derby car, i.e., build it for them. They lost anyway.

Ann's graduation from Balboa High also looms large in my mind.

We were ready to leave for the evening ceremony when Joe discovered that the flash of his office-loaned camera didn't work.

Joe's lack of mechanical expertise extended to cameras. It wasn't so much the camera itself that bothered him—he was a good photographer—it was the flash. It wouldn't go off. Either that or it wasn't correctly connected to the camera, and no pictures resulted.

Joe picked up the phone and called around but no one had the answer. Next, Joe closed the bedroom door and really got down to business. He called CIA headquarters in Langley, Virginia and spoke to the duty officer manning the hot line. The hot line was reserved for emergencies—rioting in the streets, threats against American lives, impending catastrophes. This was an emergency, of sorts, with Ann seated near the front door, tears rolling down her cheeks.

"Kirby speaking."

"Good evening, Kirby, this is Joe Kiyonaga calling from Panama."

"Yes, sir, Mr. Kiyonaga."

"Do you know how in the hell to operate a Canon F1? I can't get the flash to work and I'm in a hurry."

Kirby barely managed to hide his surprise.

Luckily, Kirby was familiar with the camera. He'd probably borrowed one from the office, too. Maybe that's one way to spot CIA operatives worldwide: they're all sporting the same model camera. Kirby gave Joe detailed instructions over the hot line, and the flash worked. When Joe opened the bedroom door, he was all smiles.

I heard this anecdote a few years ago at a party—from the man who had been on duty that night. Joe hadn't told me.

The man's name is not Kirby and the camera is not a Canon.

❧

IT STARTED WITH STOMACH CRAMPS, AFTER AN ARGENTINE-style barbecue. After a checkup, Joe was admitted to Gorgas Hospital, a United States Army facility in the Canal Zone, in January 1973. The doctors had found a stomach obstruction, but they felt that an operation would correct it. I was afraid it was Joe's ulcer acting up again.

Joe developed pneumonia during his hospital stay so the operation was delayed. Two days before his operation he was promoted to GS-16, the rough equivalent of a brigadier general.

He underwent his operation on February 17 at 8:00 A.M. The procedure was to take about two hours.

I was alone in the waiting room. I preferred it that way. I insisted that Ann and Paul attend school. There was no sense in making too much of the operation.

❧

YOU CANNOT SEE THE CANAL QUAYS FROM THE WAITING ROOM of Balboa's Gorgas Hospital, but in the Panama sky wheel hundreds of birds—gulls, cormorants, kites, hawks and buzzards. I watched them from the window until I saw the doctor coming toward me, his green operating gown wrinkled from four hours of intensive work, nose mask dangling from one ear.

With immense weariness the doctor said,

"Mrs. Kiyonaga, it's bad."

I would've preferred that he had taken me aside to talk to me. That hall was just too public. People whom I felt that I had almost come to know during our mutual waiting-room vigil perked up their ears. Misery loves company and I had none. No family, no friends. For the first time I was facing bad news with no Joe beside me. The doctor was a good surgeon, but no diplomat.

"I had to remove three-quarters of your husband's stomach. I'm quite sure it's cancer. It may have spread to the lymph nodes."

I steadied myself against the door jamb. The doctor didn't seem to notice.

"I suspect it's in the lungs, too."

My knuckles whitened against the painted wall.

"But, doctor, isn't that combination fatal?"

Only then did his eyes, hollowed by fatigue, meet mine.

"All cancer is fatal, Mrs. Kiyonaga."

✍

I DON'T RECALL MUCH ABOUT THE WALK TO THE DEPUTY governor's house near the hospital. I do remember the gleaming white hallways of the hospital and thinking that the air-conditioning was turned up too high.

At the deputy governor's house, I used the WATS line to call Mary in New York. I asked her to come to Panama immediately. She promised to take the next flight.

I quickly returned to the hospital, but by then Joe's condition had begun to deteriorate. The operation had been complicated by the pneumonia and now he was hemorrhaging. He was still unconscious. I stayed with him, by the bed, all afternoon and through the evening.

The doctors came in to check him at about 8:00 P.M. They'd been by many times. They looked at the monitor and then one turned to the other and suggested under his breath, "Better get a priest over here." They didn't consult with me.

It was a priest we knew, Father Kennedy, from nearby Saint Mary's. Only a few weeks earlier, we'd shared pizza and clams at the Restaurant of the Americas. Now he was tracing the sign of the cross on my husband's forehead, administering the last rites. Joe was still inert, oblivious.

His hemorrhaging stopped sometime later. I greeted the news with an eerie detachment, as if they were talking about someone else's Joe.

I stayed with him until 9:30 that evening when I was summarily dismissed by the nurse. Apparently, I was in her way. Some Army nurses are a pain in the neck. Her parting words to me were prophetic:

"You don't," she said, "have to carry on as though it's the end of the world!"

⤝

I SURPRISED MYSELF. I DROVE HOME WITHOUT AN ACCIDENT, though I was pulled over briefly by a Canal Zone police officer. He came over wearing his felt hat, looking like a Park Ranger. (They all had to be around six feet.) He knocked on my window. I just looked up with a "now what?" expression. He was very soft-spoken as he said,

"Excuse me, ma'am, you'd better turn on your headlights."

I remember that, as I rolled up my window and turned the lights on and headed out of the Canal Zone, for the first time that day, I began to cry. I just couldn't handle his kindness.

Ann and Paul met me at the door so anxious for good news. I put my arms around each of them and headed for the darkness of the terrace. I didn't speak. I didn't trust myself enough to say anything for fear I'd start crying again.

The last they'd heard, their father was doing well. I didn't care to disabuse them. I told them that Joe had gone through a serious operation but was "doing fine."

Paul wanted my help with an English composition, usually his father's job. He'd chosen to describe the view from the balcony, something about all his "feathered friends." He crouched on the stool next to me, writing diligently, as the lights of the ships blinked in the distance. I calmly made some suggestions, and had him read it back to me. Paul acted as if everything were fine. I think he knew, though.

Ann could tell I was distraught. I was just too subdued. Since Joe was allowed no visitors, other than me, she made a card for her father. It depicted a tiny bird fluttering right outside our terrace railing, chirping the words: "Some friends have been asking for you."

The evening of Joe's operation marked the beginning of many things. It was the beginning of completely different lives for both of us—the beginning of a long good-bye. For the first time I had to consider the prospect of life without my husband. That evening was also the beginning of my drinking. It didn't last, but it sure helped.

I suppose I thought it would make me relax—and, possibly, sleep. It didn't. Joe and I had never had difficulty falling asleep. He slept soundly; I was a light sleeper. I'd waken at the slightest sound, but fall back asleep just as fast.

When I was expecting Ann, and threatened to miscarry, I made a "promesa" to Our Lady. If she would spare Ann, I would recite the Rosary every night for the rest of my life. About five years of that and I'd had it. I'd be on my third Hail Mary when I'd fall asleep. Not even a Rosary would have done it now.

I met the next afternoon with Dr. Irwin, the chief of Gorgas Hospital, and Dr. Montague, the surgeon. They were well-intentioned, competent doctors, but they were protecting their positions.

They sat me down and proceeded to tell me that Joe had a type of cancer that defied treatment. It was useless to "seek cures." Joe had three months, six months at the outside. We were simply to make the most of the time remaining. Their words: just "enjoy yourselves."

I had seen the enemy.

I questioned their judgment. Certainly, some hospitals in the United States were working on his type of cancer? I'd read so much about the marvels of modern medicine, particularly as it applied to cancer. Perhaps the amazing advance of science offered some hope of a breakthrough? Dr. Montague did suggest that, if we did want to search for another hospital, we might consider a "teaching one."

Joe didn't know any of this; he was still groggy from surgery. The doctors had yet to break the news. I pleaded with them not to tell him. He'd lose his fighting spirit if he knew there was no hope. You can't do that to him—at least let him recuperate before you tell him. Give him a week, ten days. Let him get his strength back. Spare him the truth for now.

I felt I had to protect Joe. With so much left to do, so much pressure on him at the office, I thought he'd need every ounce of confidence possible. The doctors promised.

Half an hour later, I went in to see Joe. I'd last seen him when he'd gotten the last rites. Now he was awake, all sorts of tubes in his chest and arms. The doctors were already there, the backs of their white coats in a semi-circle before me. Maybe I shouldn't have been surprised. I came in as they were telling Joe he had six months left. Joe still hadn't seen me when I overheard his reply:

"Bullshit."

One day after his operation, in the midst of a transfusion, Joe Kiyonaga was still his own man. I had underestimated my husband.

※

JOE AND I TALKED ALONE. CALMLY. NO AWKWARD SILENCES. No tears. All business, with Joe calling the shots. Neither of us fell apart—out of consideration for each other.

I was to tell the older children the truth, the full prognosis. Paul, age nine, would be spared. He was only to know his father was sick, and that he should enjoy as much time with him as possible. If asked, I was to tell Paul it was a matter of years.

On my way out of the hospital, to pick up Mary at the airport, I ran into Dr. Martin Liebermann. He had been Joe's internist and took a more optimistic stand. He felt that Joe's cancer had been a long time developing; he termed it "chronic," and felt that Joe's strong constitution and natural immunity might be able to overcome the disease. Perhaps his was a simplistic approach but it came just when I needed it most. I took heart.

That night, I started the calls. First, David, then John. David was at Tulane Law, John at Georgetown. Both were in the middle of exams. I repeated exactly what the doctors had told me. I didn't soften the blow.

"Daddy's dying."

Both were very quiet. The first thing they did was try to reassure me. Already, they were instinctively practicing being the man of the house.

While Joe was still at Gorgas, Mary and I got busy. We called my brother-in-law, Dr. Frank Cyman, in Detroit. We called Mary's prospective father-in-law, a physician in New York. We called NIH, the Houston Tumor Center, and finally, Memorial Sloan-Kettering. Some encouraging news: Sloan-Kettering was conducting a program tailored to Joe's needs. Getting into the program was the problem. Joe knew nothing of any of this.

Within days, we got word from my brother-in-law—he had contacted a friend, an internist, who was on the staff at Sloan-Kettering. The friend promised to help as best he could. Later that week we got the news: Dr. Paul Sherlock, a gastrointestinal specialist at Sloan-Kettering, had agreed to accept Joe as a patient. I have Mary and Frank to thank—truly, I could not have done any of this without their help.

Joe was on his way to the best cancer hospital in the world.

❧

JOE WAS AT GORGAS HOSPITAL FOR TEN DAYS AFTER THE surgery. I'd cut out during visiting hours each day at noon to attend Mass at Saint Mary's where I'd hear Joe's name included among the "prayers for the sick."

I came home with Joe from the hospital on a Thursday. Owen, our driver, drove our electric blue Chevrolet Impala with obvious caution. I almost wished he'd sped up a little. Joe, in his gray flannel suit, stared forward, as if seeing the world afresh.

Joe walked into our apartment fifty pounds lighter than when he had walked out. His mustache, a little droopier now, had started to gray. All the kids were assembled on the terrace talking and ready to greet him. He walked with me straight on past to our room and, without a word, took me to bed.

Believe me, Joe wasn't dead by a long shot.

The next days were a painful progression of half-eaten omelettes and bland casseroles. *Sancocho* didn't even appeal to him. Every meal was a trial. Would he keep it down? Would he start vomiting blood again? Each trip to the bathroom scale was a heart-stopper.

All of this was especially hard on Paul. He couldn't sleep. I'd be dozing, at about 3:00 A.M., when Paul would appear at the foot of my bed.

"Mom, I can't sleep."

I'd try talking to him about his day, offer to get him a snack. I was getting so exhausted by this that one night I gave him a small chunk of one of my Equanil tranquilizers, prescribed by Joe's doctors to help me rest. It did the trick. It got to the point where I'd just hand him half an Equanil when he'd walk in. Getting up in the morning for school was another story. The maids would try. I'd try. I marvel that he managed to make it through the fifth grade.

❧

JOE WASN'T JUST IN PRECARIOUS HEALTH, HIS JOB WAS IN doubt. He had to show he could continue as chief of station. In just three weeks, he would be flying to Washington, D.C., with a stopoff at CIA, before heading on to New York and Sloan-Kettering. His bosses at headquarters would be watching him closely to see if he was equal to the task. Initial impressions mattered.

We started a regime that would have been the envy of Rocky

Balboa. Joe exercised, walked and sunned daily. Each morning, we would walk through the neighborhood; each day, he would make it a little farther. We had his tailor come by and alter his suits so that they wouldn't hang on him. I kept after him not to stoop.

It was a total sham. All surface. Joe was so weak, I was afraid he would be too exhausted to make the trip alone.

But he made the trip. And when he stepped off the plane in Washington, he looked like the old Joe. His CIA colleagues, on hand to greet him, were gratified, but embarrassed. They had an ambulance and stretcher waiting on the airstrip.

❧

THE AGENCY REALLY BACKED JOE UP WHEN HE WAS DOWN. When Joe held out for treatment at Sloan-Kettering, rather than Georgetown Hospital, they complied—at considerable extra expense.

When Joe asked that he be allowed to return to Panama and finish his tour, the office acquiesced—despite misgivings.

They did everything they could to support Joe, and their support made his recovery possible.

Dr. Sherlock started Joe on prophylactic doses of chemotherapy (5-fluorouracil). Nobody enjoys chemotherapy, but it kept Joe going. Within a few weeks, he returned to Panama.

Meanwhile, I had started reading Adelle Davis, a renowned nutritionist who pioneered in the field before it became respectable. Joe started taking massive doses of vitamins, especially B complex and E vitamins. I bought a vegetable liquefier and was amazed to realize the amount of nutrition that could be packed into a tumbler-sized drink. The basic ingredients, such as beets (and their greens), carrots, zucchini, and raw liver, I would doctor with soy sauce and lemon juice. I added fresh honey, salt, and lime juice to his fruit drinks.

But I honestly believe that our biggest break was in hearing about an obscure herb, *kalahuala*. The root of a tropical fern, *kalahuala* was reputed, locally, to cure cancer. It was introduced to us by Panamanian friends, the Noveys, who vouched for its curative powers. (Their caretaker at their vacation home had been diagnosed with liver cancer. All traces of the disease had disappeared once he took *kalahuala*.) Joe dutifully drank a tea made from it three times a day. It came to represent his lifeline, so great was our reliance on it.

Psychologically, it was indispensable. It was hard to come by. Under the best of conditions, during the rainy season when *kalahuala* flourishes, real physical labor was required to dig it from the ground. About three plants went into one day's supply of tea. Panamanian friends continually furnished us with *kalahuala* from their farms. Once we were into the dry season, the ferns dried up. Joe turned to a friend of his, a commander under Noriega, who, in turn, notified his troops in the north of Panama, the Boquete region where the climate is always moist. These guardsmen personally went out deep into the hills and dug up *kalahuala* for Joe—and shipped it to Panama City by military plane.

Joe started shadowboxing with two-pound lead weights he'd had made. At the beach in Taboga, he ventured out for a short swim. I watched him stroke slowly into the incoming waves. He didn't appear to be making any progress. I kept watching him, anxious. Still no progress. Then I saw him break through the waves and start heading for the open sea.

Joe was beginning to recover his strength.

❧

WE WERE NEVER COMPLETELY HONEST WITH EACH OTHER about Joe's prospects. We couldn't let ourselves be. We worked hard to protect each other.

We'd always worked together, with me in a support role. But now we were a true team, with barely a hint of dissension. Joe didn't object when I suggested that he get his suits altered, walk another block, sit out in the sun, or quaff another liver and beet concoction. I didn't object when he had me up in the middle of the night boiling tea or running around looking for lead weights.

We were partners in the fight of our lives. Cancer brought out the best in both of us.

≫⊱

JOE AND I WERE ENJOYING A BULLSHOT ON OUR TERRACE IN Panama one Saturday. That view of the sea was fascinating, constantly changing. We watched the ever-present line of ships on the horizon waiting to enter the Panama Canal. Joe likened our view to that of Kowloon Bay in Hong Kong. I doubt that Hong Kong could ever have been as exciting as Panama was for me that day.

I can just see Joe. He was wearing his at-home uniform—white ducks (bought at the Navy PX for $7.00), the hem of his boxer underpants clearly visible, faded navy jersey and bare feet.

Joe covered my hand with his.

"Bina, I adore you."

It had taken twenty-five years.

≫⊱

EVERYTHING IN OUR LIVES HAD CHANGED, BUT THERE WAS ONE constant: Joe's work. Late in our tour, Dave P., Joe's boss at headquarters, paid us a three-day visit.

The office organized a beach party to honor Dave, a carefully planned potluck affair with recipes attached to each pot. There was even a lighting committee (chiefly paper bags filled with sand and a candle) and a music (very important) committee. Everything was provided for, except the weather. Party day arrived and it poured.

Joe quickly decided to hold the party in the office itself. Since the Agency was conducting tours of CIA headquarters at home, he saw no reason why we couldn't have a mini-tour and party of our own. It was a stroke of genius. What would have been just another beach party turned into a "first" for all of the station wives and their husbands.

Louisa (Joe's secretary) and I had two hours to convert a collection of austere windowless offices into an Italian garden. We hauled plants, small trees even, flower arrangements, turquoise blue tablecloths and silver candelabras to the office. The reception area, which housed three secretaries, became the bar, with the stereo positioned atop the filing cabinets. Joe's office, awash in Colombian orchids (thanks to Gloria), housed the buffet table (aka desk). Joe's deputy's office we converted into an after-dinner cafe. The flickering candles everywhere were a little spooky (very apt) but effective.

During dinner, Joe proposed a toast to Dave welcoming him to Panama, and more particularly, to the office. Dave responded with a gracious pep talk, acknowledging the valuable work being conducted by the station and the importance of wives in their overall effort. We were no longer an adjunct. We were part of the team.

With that, groups of husbands and wives were started on their tours. The technical staff, housed upstairs, each gave a short talk on their duties—photography, surveillance, radio interceptions, etc.—and showed us how the highly sophisticated equipment worked. All sensitive materials were out of view.

One door remained closed, with a large DO NOT ENTER sign. It was the communications room, although not identified as such. Openness did not extend to communication.

I was fascinated. I saw offices that I didn't know existed. In my twenty-five years as a CIA wife, I'd never seen anything but my husband's office and the reception area.

Drinks were downed, candles gutted, and we danced until two. It was a great party, and a respite from everything for Joe and me. We all felt renewed pride in CIA's world role, especially since we were all made to feel that we'd contributed to its success.

We were driving through the nearly empty Canal Zone (whizzing, really) when I heard a siren. It was an MP and he motioned us over to the curb. Joe was affronted. He had diplomatic plates—not to mention that his boss was in the car.

"Let me see your license, sir."

The MP was polite.

That's more than I can say for Joe. He simply took off. (More evasive tactics!) I glanced back and the poor kid was standing in the middle of the road staring after us, too stupefied to give chase.

Once in the privacy of our bedroom I remarked to Joe.

"I gather you forgot your license?"

"That's right."

❧

OUR FOUR-YEAR TOUR WAS ABOUT UP. JOE CABLED HEADquarters asking for an extension. We were both superstitious. Joe was looking and feeling like himself—and we didn't want to leave *kalahuala*.

Headquarters sent a negative reply. Someone had already been appointed to succeed Joe. (I gather that people were waiting in line.) But they did send him a counteroffer.

In 1975, two years to the day after his operation, Joe was notified that he was to go to Brazil—as chief of station.

❧

LIFE IS FUNNY. JUST AS WE WERE LEAVING PANAMA, THE Church Committee of the U.S. Congress began its investigation into alleged CIA wrongdoings. Here was Joe, living—and giving—his life in a silent war, while the Agency's efforts were being decried at home. It seemed as if almost every day the headlines in *The New York Times* or the *Village Voice* would scream news of the latest CIA "misadventure." From here on out, a whole generation would think of the CIA as synonymous with "dirty tricks."

I think it was right then that Joe decided he wanted to stand up and be counted.

❧

SOME YEARS LATER, OUR SON JOHN WAS LUNCHING WITH
some fellow law associates in Manhattan. He mentioned that his
father had been with the CIA. "Sally," a young attorney, was seated
across the table from John. She narrowed her eyes and asked incredu-
lously, "and you're proud of that?"

John assured her that he was.

He realized that he was dealing with a child of the sixties, that
her reaction was understandable. But he wasn't about to apologize
for the CIA.

He didn't need to.

He knew his father.

CHAPTER XVII

❧

BEGUINE AGAIN

JOE WAS VOMITING.

He was four rooms away, but there was no disguising the sound. Half retch, half sob. In the middle of an overseas phone call I heard the beginning of the end . . . and I was powerless to hang up the phone.

Mary was calling from New York with great news. She had been selected to appear on the *$20,000 Pyramid*. She was going to make "big bucks," and all our worries would be over.

Rather than alarm her, since a call from New York to Brasília was a rarity, I tried to maintain a semblance of normalcy.

Joe was vomiting. Mary and I were trading banalities.

Mary asked to speak with her father. I lied. We were entertaining and Daddy couldn't leave our guests. I promised to call her back the next day.

How could the study look so peaceful? Why didn't the pictures fall off the walls—rows and rows of meticulously hung pictures that chronicled our twenty-eight years of marriage and the happenings of our five children? How is it that my flower arrangement, which I'd so lovingly prepared that afternoon, didn't collapse (or at least wilt a little!)? How dare those spindly petunias and moth-eaten baby's breaths (both a product of our Brasília garden, courtesy of Mr. Burpee) look so happy? Even the bamboo, usually pretty uncooperative once cut, remained green and proud. It was insulting.

Why did I bother to straighten my black skirt as I stood up? Why tuck in the white silk shirt? (Even then I was practicing being in mourning). I surprised myself. Was it possible that I was finally attaining the unattainable? Was Bina, the hysterical, going to be able to face up to a crisis—the worst crisis of her life—without losing control? Here was my chance. I ran from the study through the dining room (its twenty by thirty feet dimensions took on gargantuan proportions) through the hall and into the guest bathroom.

This bathroom was my particular pride. I'd transformed it from a walk-in water closet to a powder blue and silver gem. Blue carpeting and walls mounted to a silver ceiling. The sink was white porcelain wreathed with corn flowers. Silver-plated dolphins spouted water. Even your reflection mirrored on the wall was rimmed with silver, a gift from my mother.

The bathroom was no longer a jewel. It was a swill pail. Joe was bent over the toilet, but too late. He'd vomited up days and weeks of undigested food all over the walls and rug.

"Couldn't you even manage to hit the toilet?" I shrieked.

I was back in character.

❧

I DON'T THINK I COULD EVER GO BACK TO BRASÍLIA.

The city is unreal—a man-made vision of utopia. The buildings are poised under the boundless blue of the sky like ikebana arrangements of twisted steel, stained concrete, mirrored glass, chrome and, even, water. Brasília puts me in mind of a sci-fi movie set plunked down in the middle of nowhere—a giant, nearly empty, World's Fair Emporium on the Texas plain. The quiet heat and yawning space make every day there seem like a lazy Sunday afternoon. It's unfulfilled promise nearly breaks your heart.

But what could compete with that glorious sky? It comes right down to the red-clay soil. You look out at the vast stretches of virgin *mato* brushland and can actually imagine that dinosaurs once walked the earth.

It was just Joe, Paul and me, and, of course, Jack the poodle. Paul had returned to the country of his birth—and loved it. He would take long hikes out into the *mato* surrounding our house (tracking back the red clay across our off-white carpeting). He took to shooting anthills with the .22-caliber rifle Joe had given him. It was his Molokai.

We'd inherited from Joe's predecessor a lovely white stucco, California ranch-style house with a red tile roof. Its row of French doors opened out onto a spacious red brick terrace and sparkling swimming pool. (CIA does things up right.) We had an unobstructed view, across our badminton court/soccer field, of Lake Paranoá (or "Lake Paranoia," as Joe used to call it) and the Brasília skyline beyond. The house, with its billowing yellow chiffon curtains and pale blue furnishings, had the feel of a sailboat moored in the desert.

We could've had such a good time.

⸙

JOE ARRIVED TIRED AND NEVER SEEMED TO CATCH UP. HE WAS plagued by colds and flus. He and Paul would jog the perimeter of our garden each night to stay in shape, with Jack yapping at their heels. (Jack was never really clear on the concept of being a watchdog.) Maria, our cook, conjured up *canjas* (chicken stews) brewed with obscure Amazonian roots and vegetables grown right in our garden.

Kalahuala was what we really needed. Soon after our arrival, we renewed our search. Ambassador Castro Alves, of the Brazilian foreign ministry, was our neighbor and a noted herbologist. He went to unbelievable lengths to secure the herb for us, calling

Rio, São Paulo and Santa Catarina in southern Brazil. He sent letters off to Lima, Peru, and Bogotá, Colombia—everywhere he had contacts. He kept us current as his search narrowed.

Maria was amazing. I have never known anyone who was so unfailingly cheerful. She hailed from the north of Brazil, Bahia, and was a Brazilian version of Aunt Jemima. She was also an adherent of *macumba*, a native voodoo-type religion. Though she never admitted to it, the evidence was unmistakable. Every full moon, Maria would leave an offering for the gods on the road behind her room. It consisted of a white, lighted candle, a freshly plucked chicken (ours, no doubt), a bottle of Brahma Chopp beer and a pack of Hollywood cigarettes. The offering was accompanied by unusual activity within Maria's room, which emanated in the form of chants, or wails and pungent wafts of incense. I was relieved when I discovered, after some checking, that her use of a white candle boded good. A black one boded ill for the future. I kept my eye on that candle.

※

BRASÍLIA IS A MAGNIFICENT CITY, BUT IT'S JUST TOO FAR FROM the rest of Brazil. Lúcio Costa and Oscar Niemeyer, Brazil's foremost architects, designed Brasília in the shape of a plane, a fitting motif as Brazilians in the city spend most of their time trying to leave. The Government Ministries buildings, lined up like dominoes on either side of a ten-lane highway, make up the fuselage; the President's Palace and Congress buildings comprise the nose of the plane. The "wings" are the rows of Soviet-style apartment dwellings, each identified by a quadrant number.

Whoever decided where to locate this new capital city should have split the difference and situated it only halfway inland. (The old capital was Rio.) Its location seemed almost cruel. Brazilians are beach people, and the beach closest to Brasília is six-hundred miles away.

Brazilians tell the story of two Bedouins, traveling by camel across the Sahara, who spied a sun-bronzed Brazilian in swim trunks jogging across the shifting sand.

Bewildered, one Bedouin shouted,

"What're you doing out here? There's no water for a thousand miles."

The Brazilian replied,

"Yeah, but isn't it a great beach?"

⁂

LIVING IN BRASÍLIA COULD BE PRETTY ISOLATING. IT'S INSULAR feel was Hawaii all over again. Little things assumed critical importance—the availability of corn flakes at the commissary or the week's movie selection at the Embassy (usually some Charles Bronson feature). Joe even had the Agency send us our own movie projector to use at home. We'd rewatch *Sanford & Son* and *The Carol Burnett Show* episodes shipped down from the States.

The Brazil we had returned to could no longer be classed an underdeveloped nation; it was a power. The country was entering into its nuclear phase, and its principal cities now boasted the inevitable badge of progress—smog. One thing hadn't changed— the Brazilians. They were still the most fun of any people on earth.

The Soviets and Chinese had established a real presence in Brazil. In fact, the Soviet Embassy, with its massive red and white facade, sat right across the street from our embassy. You almost got the feeling they were trying to stare us down.

Joe got together one evening with his Soviet counterpart, the head of the KGB in Brazil. The meeting was Joe's idea. (Maybe he was trying to finally land a defector.) Despite their competing ideologies and objectives, they did manage to find one area of common ground: martinis and Westerns. Joe set up our projector, and they watched a Western featuring the KGB man's favorite star, Chuck Connors. Joe switched to vodka martinis for the occasion.

Work can be a salve. It invigorated Joe. But with so much

happening in Brazil, it's unfortunate that Joe's last posting was his worst in terms of personnel. The office was disorganized, morale terrible.

Agent recruitments were way down when Joe arrived, despite the creative efforts of one operative, a talented Chinese-language specialist (versed in three dialects) who had been sent to Brazil to penetrate the Red Chinese community. (I recall Joe's telling comment to me: "It's not the Russians that worry me. It's the Chinese.")

Realizing that a man's stomach often controls his heart, and head, this operative had begun to frequent Chinese restaurants intending to meet his Chinese counterparts, and eventually to befriend them. He ended up befriending every Chinese chef and waiter (and waitress) within driving distance of Brasília. He became a walking Michelin Guide on local Chinese cuisine. He would salivate when describing the various dishes for which each house was noted. He was gaining weight but he had yet to gain one agent.

Part of the problem was that recruitment had gotten a lot tougher. In the mid-1970s, the CIA was still getting a battering in the press worldwide, which made potential recruits chary of doing business with the Agency. The Church Committee (and, to a lesser degree, Bill Colby) were busily pulling the rug out from under the Agency—especially, the agents in the field. (Joe had always admired Bill Colby, but he was disturbed by the revelations he was making to the Church Committee. Joe felt that Bill perhaps didn't truly realize to what extent he was harming his buddies in the clandestine services.) Joe was doing his best to shore up the damage in Brazil.

A priest agent of Joe's, from our São Paulo days, contacted Joe, in a state of great alarm. Joe immediately flew to São Paulo to calm him. Father was afraid that his name would appear in the U.S. newspapers. He threatened suicide. Joe assured him that his name had been wiped clean, without a trace.

I suspect that the really devastating blow to the Agency came in an exposé written by a disaffected former CIA operative—Philip

Agee. My personal view is that he may have been responsible for
the death of a friend.

※

JOE HAD KEPT IN TOUCH OVER THE YEARS WITH DICK WELCH.
Dick had been sent to Greece as chief of station at about the
same time that we'd gone to Brasília. At that time, chiefs worldwide generally had the same pretty impressive cover title, a fact
duly noted in Agee's book. After Dick arrived in Greece, an anti-
CIA publication with suspected KGB ties published Dick's name
and CIA affiliation.

It was a few days after Christmas 1975 that Joe got a letter
from Dick. I remember him reading it aloud to me. It was long,
newsy, full of folksy quips. Dick was looking forward to a quiet
tour; it looked as if there'd be lots of time for sailing. Greece had
the same feel as Guatemala as far as work was concerned—"lots
of chasing of Indians up and down dark alleys."

The timing of the letter could not have been more eerie. Dick
Welch had been gunned down by an assassin five days earlier—
the night before Christmas Eve.

※

JOE'S INTERNAL OFFICE SITUATION SOON BEGAN TO IMPROVE;
there was less friction, more cooperation, and more production.
He traveled across the country visiting the various bases and managed to zero in on three targets for recruitment. Monthly intelligence reports went up (a mark of an active station), many for
dissemination to other agencies, which is a singular honor.

Joe arrived home one evening with the makings for *caipirinhas*,
and the ingredients for a *churrasco*—spareribs, sausages, pork loin
and beef fillet. He'd received a letter of commendation from
headquarters.

The work energized Joe, but he still seemed tired. He started
getting headaches, which he'd never had. At my urging, he consulted

a stomach specialist in Brasília. They hospitalized him and performed gastroscopes. The result was negative; no sign of a tumor. They did complete blood work. Everything was normal. They X-rayed his chest—again negative. Finally he checked with the Embassy doctor who prescribed Valium. His findings were that Joe was "home free" as far as cancer was concerned. I wasn't so sure.

❧

WHILE WE WERE IN BRASÍLIA, JOE FLEW UP FOR THE ANNUAL consultation at "the Farm." Every year, the various chiefs of station are brought home to meet with each other at the Farm, and then head on to headquarters for meetings with their opposite numbers at Langley. It's a great chance for the chiefs to touch base, renew friendships, mend fences and make deals. Joe always enjoyed this yearly trip, as much as he dreaded my shopping list that went along with it.

Quarters at the Farm are comfortable but not lavish. The evenings are a time to unwind over drinks before a man-tailored dinner such as steak, baked potato, and salad, followed by cognac. The cooking is good without being fancy. A different movie is shown each night. One evening Joe suggested to a friend that they see *Deep Thrust* (a kung fu disaster). The friend accepted thinking that Joe had said *Deep Throat*.

Joe was at headquarters and heading for a meeting when a friend stopped him in the hall. He mentioned that Jack Downey was in a nearby office being debriefed. Would Joe like to speak to him?

You can just imagine how Joe felt. Did he feel like speaking to the man whom he'd sent on a failed mission twenty-one years ago? Who'd languished in solitary for many of those years? More to the point, would Jack want to speak to Joe?

The two men met. No words were necessary. Their firm embrace said it all.

❧

CARNAVAL IN RIO WAS A GIFT WE ALL GAVE EACH OTHER, BUT
that none of us really wanted.

I questioned whether Joe was up to it. The dancing till dawn,
streets alive at every hour with roving samba bands? He assured
me he was.

Our day would begin about four in the afternoon, when we'd
jostle through the riotous crowds and samba bands along Copacabana's Avenida Atlantico. People didn't walk, they danced. Many
were in costume. Most were drunk, but Brazilians are good-
natured drunks. Drinking just makes them that much more fun.
We'd pick up some chilled coconut milk from a sidewalk vendor
and eventually make our way to our favorite hotel, the Ouro
Verde, for an aperitif and quail eggs.

Most of Carnaval centers around the judging of the parading
samba schools. It's a three-night affair starting at midnight and ending after dawn. I wish I'd paid more attention as the distant roll of
a thousand drums announced the approaching "Padre Miguel"
School. The drums were attached to a thousand clean-shaven, white-
tuniced men—the "monks" of Padre Miguel. Three hundred yards
away, they picked up the beat and by the time they arrived at the
reviewing stand the roll had become a roar, in perfect unison. The
crowd samba-ed to its feet. An unbelievable spectacle.

I nearly missed it all. I was too busy watching Joe struggle to
his feet, trying to rise to the occasion. Too busy watching Paul
watch his father.

We'd arrive back at our Agency-funded apartment, situated
beautifully on Copacabana Beach, at five or six o'clock and go
for a swim in the ocean as the sun was rising. Maria, whom we'd
sent on from Brasília, would serve us a breakfast of papaya,
mango, pineapple, omelettes, ham, croissants and thick Brazilian
coffee that she strained through a silk sock. We'd bed down and
start the rounds all over again at four that afternoon.

It was our last good time in Brazil.

❧

CARNAVAL HAD BEEN JOE'S IDEA. SO WAS PLAYING IN THE AN-
nual Embassy tennis doubles tournament. He put himself through
all this not for me, not to convince himself of anything. He did
it for Paul.

I attended their final match. Paul scurried all over the court,
while Joe struggled to get to the ball. They were getting creamed.

Joe was so weak. I could see it in his eyes. But he didn't let
on—no hint of pain, no betrayal of discomfort. With each rally,
as he willed the ball over the net, I didn't know whether to yell
for him to please stop or stand up and cheer.

It was that next weekend that the vomiting started.

I'd had it. I didn't care what the local doctor had to say. I
knew Joe was dying. I gave him an ultimatum. Unless he left on
Pan Am's Friday flight to New York (it was Wednesday), I would
pack up and leave, taking Paul with me. It was the only time I
had ever threatened to leave him.

Joe refused. I insisted. We argued, we talked, and, finally, we
bargained. At 3:30 in the morning, he agreed.

Joe arrived at Sloan-Kettering and was immediately hospitalized. The doctors sent for me. They were going to operate. We
left Paul in the care of the Chinese cuisine expert.

I arrived in New York in mid-April. Spring was barely noticeable in the city, but it was everywhere in the hospital. The lobby
was a garden of azaleas, tulips and forsythia. Whenever I think
of Sloan-Kettering I think of flowers, great doctors—and Joe.

Joe greeted me at the elevator. He looked fit and tanned—
very out of place with the pale, bent patients whom I saw wandering the halls.

He had a guest, a coworker from the CIA who'd flown up to
appraise Joe's situation and to be briefed by Joe on the Brazilian
scene. It was natural that CIA should be concerned about Joe.
Apart from their compassion, they had to plan ahead. Brazil was
and is a critical post. The station was shaping up and needed a
station chief.

I met Joe's doctor, Paul Sherlock, and his surgeon, Michael
Paglia. I liked them and their approach to Joe's illness. They were

realistic, but optimistic. I felt that if anyone could save Joe, they could.

Sloan-Kettering is strictly a cancer hospital. The personnel is geared to cancer—and death. Unlike hospices, they refuse to accept the finality of cancer. A sense of adventure, and hope, prevails. The doctor encourages the patient, and one patient encourages another.

⊱

DR. PAGLIA OPERATED AT EIGHT IN THE MORNING ON THE Thursday before Easter.

I waited in the main lobby of Sloan-Kettering. A crew was at work arranging and trimming the flowers and plants that were everywhere throughout the lobby.

I recall seeing a young woman being wheeled into the large room. She was unusually pretty and was wearing a pale blue pleated negligee trimmed with blue satin. Her family was waiting for her—her husband, and three small children. I could see that this was the first time that she'd seen her children since her confinement, and the springlike atmosphere seemed to help. Everywhere around us the scene was repeated.

At eleven, I was called to the receptionist's desk. Dr. Paglia was on the phone, calling from the operating room.

"Mrs. Kiyonaga, I have some bad news and some good news." (I've always hated that expression.)

"First the bad. We found a tumor in the stomach and a growth in the groin. I was unable to remove the tumor. It's too large. But I was able to do a stomach bypass. Dr. Bazell (he was the urologist) is entering the operating theater. He'll try to remove the tumor in Mr. Kiyonaga's groin. I'll keep in touch."

Half an hour later, a nurse on the surgical staff came to speak with us. Joe was tolerating the surgery well and we would hear from Dr. Paglia within the hour.

I was astounded. In the midst of a grueling surgical procedure the doctors—and nurses—of the staff had enough compassion to think of the patient's family, anxious for news.

❧

JOE DIDN'T STAND A CHANCE. THE DOCTORS KNEW IT; JOE knew it; I knew it; and so did the children. Stomach cancer is an especially devastating type of cancer. Sloan-Kettering may be the best that the world has to offer in the way of cancer treatment, but the best was not going to be enough.

I found myself practicing at being a widow—and trying to spot other widows. I would visit the hospital library on the seventh floor. Most of the volunteers were women in their sixties and seventies. Probably many of them were widows. In turn, each of them would disappoint me with mention of "my husband this" or "Peter that." I became jealous.

My best hope was the social worker on Joe's floor. She was a trim, good-looking woman, but a little down in the mouth. She had to be a widow. Finally I asked her if she had a husband.

"Yes. He's a writer."

She not only had a husband—she had a halfway interesting one. My jealousy turned to resentment.

❧

"I'D LIKE TO SEE ONE SMILING WIDOW."

I was speaking with Father Quiñónez outside Joe's room.

"I'm smiling, and I'm a widow."

I turned to face a member of the medical staff. About my age, she put me in mind of me. Her white coat distinguished her. Otherwise she was a middle-aged woman with short hair, a kind face—and she was smiling.

At last I'd found my counterpart.

She reminded me of someone else I'd known.

❧

WHEN I WAS IN HIGH SCHOOL—AT TRINITY PREP—I MADE
friends with a boarder, Patricia Logan from Cohasset, Massachusetts.
Pat was shy, perhaps because she stuttered. Believe me, I had no
idea when I met Pat that she was part of some sort of dynasty. I
just liked her. She had such an open, Irish face—and I wanted to
be her friend. I invited her to my house on Harlem Avenue for the
weekend and she loved it.

That summer my folks received a letter from a "Mrs. Edward
Logan" inviting me to spend a few weeks with Pat's family. Delighted, I took off and was met at the Boston Airport by Pat and her
brother, Eddie. I was ready for an adventure, but nothing could prepare me for the Logan home on Jerusalem Road in Cohasset. To my
untrained eye, it looked like a baronial estate.

What surprised me was that, on the one hand, I was faced with
this imposing scene while, on the other hand, there was Pat—really
unassuming. Those girls back at Trinity had no idea what they were
missing.

Even more surprising was the cocktail gathering that greeted me
in the Logans' living room. I remember meeting the renowned Mayor
James Michael Curley and Sen. Leverett Saltonstall. And Pat's
mother—a truly imposing figure, whose perfectly cultivated tan set off
her beautifully coiffed silver-white hair. She was dressed completely
in black, from her black designer suit to her sheer black hose.

Mrs. Logan was every bit as impressive in her politicking. She
very much wanted to have a public edifice named after her late
husband—Gen. Edward Logan, the storied founder of the 26th Division, the Yankee Division that had fought with distinction during
World War I. She was an independent woman—a widow—thinking
on her feet. I liked that. And she was smiling.

Her lobbying efforts paid off. She's why the airport in Boston is
called Logan.

✎

JOE'S SURGERY WAS FOLLOWED BY COBALT, AND COBALT BY
massive doses of a new type of chemotherapy. He had asked

to return to Brazil and receive chemotherapy there. A doctor was sent to New York from CIA headquarters to assess Joe's situation. You couldn't fool that doctor, though. One look at Joe and he realized that Joe would not be able to work again— at least not under the taxing conditions that the Brasília job imposed.

It was decided that we would go back to Brasília for a six-week rest, and then pack up and return home.

When we arrived in Brasília, Paul was waiting for us at the airport, watching us through the glass partition at the baggage claim. Joe, so thin, couldn't lift his duffel bag off the conveyer belt. Paul understood now.

Joe swam and sunned every day, a repeat of his Panama regimen. The other children had flown in and ours was a charmed circle. I began serving him a liquid diet at home—more chicken *canjas*, vegetable drinks, custards and omelettes. I cautioned him to eat slowly. I even made a small placard, hand decorated with tiny flowers, that I propped up in front of his plate. It read: PLEASE, JOE, EAT SLOWLY. I was killing him with kindness at the same time that I was nagging him to death.

Each night after dinner, I'd watch Joe and the kids head off on their ritual after-dinner stroll. They'd walk, five abreast, up the street from our cul-de-sac, talking and laughing. From my vantage point, they looked like such a healthy, carefree family. I'd almost forget.

Each day was a dividend.

The first shipment of *kalahuala* arrived two weeks before we were set to leave. Ambassador Castro Alves had been searching for months. He'd finally located some in Colombia. He hadn't let us down.

Joe had unfinished business: he wanted to recruit the three targets he'd identified. He arranged to lunch with them on three successive days and recruited each of them in turn—his farewell gift to the Agency.

AMERICA WAS CELEBRATING ITS BICENTENNIAL THAT SUMMER.
We celebrated our Bicentennial Fourth of July with the British
Ambassador and Lady Dodson.

She, Julie, had first endeared herself to me when I heard that
once, while eating at a neighborhood restaurant, she'd asked the
management to turn down the air-conditioning. They refused.
The ambassador's lady retaliated by going from table to table and
draping every available napkin on her head, shoulders and bosom,
before regally ordering her gin and bitters.

When we arrived, their butler showed us from the driveway,
past their dove-gray Bentley, and into the house where the ambassador and his wife waited in dinner jacket and long dress. Behind
them was a bucket with Dom Perignon on ice and, beside it, a
tray of hors d'oeuvres. When the bottle was finished, Joe mixed
a pitcher of martinis and we settled down to serious drinking.
Whatever its drawbacks, drinking can be more than a tonic and
that evening showed it. It was our second favorite farewell, after
our curbside martinis on Leland Street.

We prepared to leave for the States.

❧

JOE, QUIETLY ARRAYED IN A STARCHED WHITE *GUAYABERA*
and bankers gray flannels, sipped his third *caipirinha*. We were
enjoying our view of the lake by sunset. It was our last evening
in Brasília.

The stark federal buildings on the other side of the lake practically stood at attention as the Brazilian flag was lowered from
the mammoth flagpole that graces the President's Palace grounds.
Sunset is an eerie time. Joe always loved it.

"When I was a kid," he said, "I remember thinking that anything was possible at dusk. I could run faster, jump higher and
whistle louder."

Sunset in Brasília takes on a magical quality. The thinness and
purity of the air and the sharp orangeness of the Brazilian sun

as it drops out of sight lend an air of luminescence—a sense of impermanence that matched my mood.

Euclides, the gardener, was cutting the grass. The rhythm of his scythe competed with the chorus of the cicadas scattered through the high grass. Roberto Carlos's "Detalhes" was playing through the outdoor speakers. Jack, our wandering poodle, loped back from the brushland, where he'd probably hidden another treasure—Paul's Frisbees were his favorites. Maria rounded the corner of the house. She was wearing her favorite uniform—hot pink-and-white print voile with a stiff white apron and bandanna. (I never saw her without one.) She was headed for the vegetable garden, on the water's edge, to pick part of the evening's meal—collard greens, parsley, romaine, endive, tomatoes and a papaya, for breakfast. She sang as she walked, something plaintive, part of the *macumba* rite.

The guard came around the other corner of the house. His khaki-brown uniform and holstered gun gave him a snappy appearance—until you made out his amiable features beneath the peaked cap. He addressed a salute to Joe. Joe did not return it. Instead, he got to his feet and extended his hand for a farewell handshake.

The pool lights came on, synchronized with the wrought-iron lights that dotted the trees surrounding the grounds. They illuminated the badminton court and soccer field—the rock flower garden that was to have been my pride. The morning glories and delphiniums were yet to bloom.

Joe shifted in his wicker chair. (Jack shifted beneath it.) He leaned forward to steady my hammock. I knew that expression—he had good news.

"Bina, I got promoted to GS-17 today."

I couldn't answer. Why hadn't it come through when it might have done some good?

" 'Ted' cabled me. I'm being considered for a good job at home."

Maybe I was wrong. Maybe the timing was just right.

As a nighttime hush fell on the terrace, Jovi parted the sliding doors that led to the dining room. She wore her formal black uniform reserved for special occasions. Beside her stood Ferreira, our driver cum butler, in his dinner jacket. Behind them, the dining room glowed. Candles, silver, crystal added to the solemnity of the occasion.

Joe and I were celebrating our twenty-ninth anniversary.

❧

CHAPTER XVIII

A LIGHT GOES OUT

THERE WAS SOMETHING PREORDAINED ABOUT THE WHOLE DAY.

It started with a call from Joe's nurse at Sloan-Kettering, before I could make my usual 9:00 o'clock call to him.

"Mrs. Kiyonaga, Mr. Kiyonaga would like to speak with you."

My husband's voice was strong but anxious.

"Bina, do you think that you could come early today?"

"I'll try, Joe, but the guards might not let me in. Are you all right?"

"I'm fine, but noon seems a long way off."

"I'll leave right away. Don't worry, I'll be there soon, Joe."

I dressed with unusual care, all in black. I combed and recombed my short, unstyled hair. Everything seemed to hold my attention. The icebox needed to be tidied up; the trash needed to be emptied.

I stepped out onto East Eighty-eighth Street and ran smack into spring. Snow was on the ground, but the air was alive with thaw. It was March 8, 1977.

Instead of hailing a cab I decided to take the Lexington Avenue subway two blocks away, under Gimbel's.

Along the way, I stopped to buy some stockings at a discount hosiery store. In the middle of my transaction, while I was actually pulling some dollar bills out of my wallet, everything snapped into focus. What was I doing? Joe was dying and I was buying stockings. This was to be the biggest day of our lives together and I was going to be late. I'd had five years to get ready; Joe had issued me a special invitation and I was going to be late . . .

❧

"COME ON, BINA."

We were strolling the Ala Wai Canal on the way to Waikiki.

"I'll take you to the Royal Hawaiian afterward." The Royal Hawaiian had about the best restaurant in Honolulu. We could scarcely afford it.

We got to Waikiki and the beach was packed, mainly with elderly couples. Deeply tanned, almost leathery, they cast a pall over the scene. Joe took it all in, and turned to me.

"Bina, let's be on the beach now."

We always were.

❧

I LEFT THE SHOP WITHOUT THE STOCKINGS AND HAILED A CAB.

"Sixty-eighth and York, please."

(Please God, I thought, please don't let this one be talkative.)

Sloan-Kettering's address meant only one thing: cancer. It evoked varying responses from cab drivers, but it never failed to elicit a reaction.

"Beautiful day. Spring's finally here."

He turned and smiled at me.

("Please keep your eyes on the road. Don't you realize I'm in a hurry? I don't have time for trivialities.")

He was Latin. They're the worst. Born psychiatrists, they can spot your mood the minute you sign on as a fare. He was going to cheer me up.

"The radio says it's going up to sixty-five degrees today."

His mood did not suit mine and I let him know it. I was polite but firm with my nod.

The hospital doorman helped me out of the cab and across the icy sidewalk. I had neglected to wear my boots.

As I rode the escalator to the lobby I closed my eyes. I couldn't bear the sight. The lobby was brimming with exquisitely nurtured tulips, hyacinths and azaleas—I almost resented their intrusion on my suffering.

I approached the receptionist's desk and was greeted by the most painfully beautiful flower arrangement I have ever seen—a single gnarled sprig of flowering quince jutting out of a tall, cylindrical glass vase. It was so spare, so Oriental—and so solitary.

"May I see my husband early today? His name is Joseph Kiyonaga and he's in room 752."

"Let me check to see if he's on the critical list . . . No, I'm sorry you can't go up. He's not listed as critical." (Here I was, actually disappointed that Joe wasn't critical.)

"Would you please call Dr. Sherlock? I'm sure that he'd let me go up."

"We can't do that, Mrs. Kiyonaga." She glanced at her watch. "It's only fifty minutes until visiting hours."

She was no help. I could tell, just by looking at her. Women in authority can be overbearing.

I approached the elevators and the guards. They knew me too well. I'd visited Sloan-Kettering every day for three months. I passed them by and walked clear through the block-square hospital to the outpatient section on Sixty-seventh Street. I would reach Joe's floor by an unguarded entrance.

Everything seemed to get in my way. The elevator was loaded with technicians, bottles and machines. The elevator stopped at every floor except the seventh. I was detoured to the eighth because of construction work. The eighth floor made no sense to me. I was faced with three doors. Beyond that I found a hall. The hall radiated in four directions; I was lost, and frustrated. I could have cried.

Fifteen minutes later I finally found my way to Joe's floor and wing.

Young, bearded doctors stood outside the nurse's station discussing a case. It was a familiar scene. The head nurse, Mrs. Ryan, passed me and smiled and I felt reassured.

I rushed down the hall to Joe's private room at the far end. His door was slightly ajar and he was wearing his glasses, watching for me. I was surprised. Joe usually had his nurse keep his door closed to insure his privacy.

My husband lay in bed; he was emaciated, a new growth of white hair barely covered his head—a man who six months earlier had been a handsome, robust Japanese-American with a head of coarse, black hair and a fierce-looking mustache.

He had suffered through surgery, cobalt and chemotherapy—and they had taken their toll.

❧

A FEW MONTHS EARLIER, I HAD READ THAT SMOKING MARI-juana offsets the bad side effect of chemotherapy—the nausea.

One day, I presented myself at a tobacconist's on Madison Avenue.

"May I help you?"

"Yes, I'm looking for a small, clay pipe."

"For what use?"

"To smoke marijuana." (I failed to explain that it was for medicinal purposes.)

The clerk hastily sold me a small pipe—actually, it was kind of cute. As he was rushing me out of the store, eyes furtively darting from side to side, I paused to inquire:

"Now where can I find some marijuana?"

❧

ENTERING JOE'S ROOM, I EMBRACED HIM AND WAS IM-pressed by the strength in his arms.

"Bina, let me go."

I understood now the early-morning summons: Joe had been waiting to ask my permission to die. He was tired and longed to rest. I knew the time had come to release him.

"Go ahead, Joe."

"Better one of us than one of the kids," Joe said, "better me than you."

Again, he was right. With my agreement, he relaxed— just as the phone rang. It was the beginning of a series of calls from Joe's colleagues to bid him good-bye. The Agency was closing ranks, saluting one of its own.

❧

THE DOCTORS CAME BY ON ROUNDS. JOE SHOOK HANDS WITH each of them in turn. The Japanese have elevated death to the status of an art form, but I didn't have to be Japanese to be impressed with his presence of mind: he was facing death with characteristic aplomb. In some respects he viewed his bout with cancer as an adventure, and hoped that his stay at Sloan-Kettering had furnished the doctors with clues in the renaissance of cancer treatment.

❧

SEVERAL WEEKS EARLIER, DR. SHERLOCK HAD CALLED ME out of Joe's room. He remarked that Joe's condition had deteriorated—there was no telling if he would live for hours or weeks. He proposed sending Joe home (to Chevy Chase) with electronic equipment and a nurse. Hospital beds were in short supply.

"But, Doctor, you yourself said that all that is keeping Joe alive is his determination and his family's support. If you send him home, he'll give up hope."

"I'm sorry, Mrs. Kiyonaga, but these are hospital rules."

"In that case, I'd appreciate a meeting with you, Dr. Paglia, Dr. Bazell and the chief of Sloan Kettering."

"Five o'clock in the conference room."

"Fine, I'll be there."

I was there all right, armed with a tiny tape recorder secreted in my purse.

The doctors I'd requested were assembled. Their white starched coats and professional demeanor nearly gave me pause—but not quite.

"Thank you for coming, doctors."

With that I took out my tape recorder, placed it on the conference table, and pushed the red RECORD *button.*

"Shall we begin the meeting?"

Joe stayed.

❧

FATHER QUIÑÓNES CAME BY AND PRAYED WITH US. HE MUST have sensed something. He asked that we join hands as we prayed, and his eyes filled with tears when he blessed Joe.

Mary and her husband and John, all living in New York, were there; Ann had flown in from Berkeley; David from Panama; and Paul from Philadelphia. Joe remarked that the Italian painters made heaven look like a good place, and that when he got there he would have a steak and a scotch. He hadn't eaten solid food in months.

Joe didn't talk long. He said he loved us and was proud of us all, that his life had been full and that the measure of how good it had been was around him in the room. His only regret? That he wouldn't live to see his children's children. No tears, no bitterness. Then he said good-bye to each of the children, and they left the room.

I stayed on. I asked Bob, the nurse, to let me be alone with my husband for a while.

Joe and I talked about many things. How much we loved each other, how proud we were of the kids, how good the years had been. We were content, happy to be together.

"Bina, can you imagine what it means to a man to know that

he's been the whole of his wife's experience? . . . And I've been completely faithful to you. I wouldn't have thought that possible thirty years ago."

He took my hand and kissed it. He pulled me toward him and whispered in my ear.

"It was good, wasn't it?"

I smiled. And then I cried.

꩜

SUDDENLY, JOE GLANCED AT THE CEILING AND AN EXPRESSION of awe and joy suffused his face. He put his left arm around me and tried to pull himself up in bed, with his right hand on the guardrail. He hadn't been able to sit up for months. He let go of the rail and reached his hand up toward the ceiling as though he were trying to grab something.

"Don't you see it, Bina? The ladder!"

"No, Joe, please lie down."

"Can't you see the light?"

"No, Joe, you'll hurt yourself." I was afraid that the electronically operated line to his heart would pull loose.

"No, Bina, it's all right. It's not as though we're criminals trying to escape from prison. This is right, it's good." He tightened his grip around me. "Come on, I can take you with me."

Terrified, I forced him to lie down and, as I did, he took a swipe at me.

"Joe, you're not going to die mad at me, are you?"

"No, Bina, you just don't understand."

Bob knocked at the door.

"I see by the chart that Mr. Kiyonaga has had an exhausting day and I'd like to ready him for bed."

As I got my things together, I reminded Joe that I would call at 10:00 (it was then 9:30) as I did every evening, to wish him good night.

"Bob, I have an idea. Why don't we return Mrs. Kiyonaga's call around 11:00 this evening?"

"Fine, Mr. Kiyonaga, whatever you'd like."

"Good-bye, Joe, God bless you. I love you."

"Good night, dear. I love you . . . always."

Joe blew me a kiss as he spoke these words and his eyes closed as I stood in the doorway and watched.

As soon as I reached my apartment, I called Joe's room. Bob said that Joe had been asleep ever since I left and suggested that he not waken him. I agreed.

At 11:10, the phone rang. It was Dr. Sherlock.

Joe was dead.

THE NEW YORK TIMES, FRIDAY, MARCH 11, 1977

JOSEPH Y. KIYONAGA, 59; EX-SOLDIER, C.I.A. AGENT

Joseph Yoshio Kiyonaga, who had served with the Central Intelligence Agency as chief of station in El Salvador, Panama and Brazil, died Tuesday at Memorial-Sloan Kettering Cancer Center. He was 59 years old.

Mr. Kiyonaga, who joined the C.I.A. in 1949, resided in Chevy Chase, Md. He was a direct descendant of the 19th-century Japanese woodcut artist of the same name. Mr. Kiyonaga was born on the island of Maui, Hawaii.

During World War II, he served in Italy and France with the 442d Regimental Combat Team, which was made up of Japanese - American volunteers from Hawaii and from internment camps on the West Coast who had petitioned President Franklin D. Roosevelt for permission to form their own detachment. The unit became one of the most-decorated in the Army. Mr. Kiyonaga was cited for valor and received a field commission.

He graduated from the University of Hawaii and received a master's degree from the Johns Hopkins School of Advanced International Studies. During his professional career he also served a number of years in Japan.

Mr. Kiyonaga is survived by his wife, the former Bina Cady; three sons, David, a lawyer in the Canal Zone; John, a student at Columbia Law School, and Paul, a student at the Hill School; two daughters, Ann, a student at the University of California at Berkeley, and Mrs. Michael DiGiacomo of Bronxville, and his mother, Mrs. Joseph Swerts of Molokai, Hawaii.

Joe and me. San Salvador, El Salvador, 1966.

Its been hard. Life without you. It's tough waking from a dream, thinking that you're there beside me, only for it to come crashing down with the realization that you're gone. (That's the worst part.) Tough listening to music on the record player—maybe "Sabor a Mi" or "Try to Remember"—and not having you to dance with. Or walking into a couples' party, and putting on a brave face with no Joe across the room to salute me before his first sip. It was even worse, that first spring, to glance out at the garden and see the single red tulip blooming beside the pond— the tulip that you planted.

But life has its moments.

Thank God for the kids. They're a reflection of you in unexpected ways: Mary with her impressive bearing; David with his disarming reserve; John with his sharp wit; Ann with her infuriating serenity; and Paul with his unfailing kindness. They're all you, Joe, and they're great.

And then there are the grandchildren, eleven of them now. You'll be pleased to learn that two of them are named Joseph, and they both look far more Japanese than any of our children. It's a shame you can't see all of this. But, then again, maybe you can.

If you can, maybe you saw me when I finally made it back to Molokai—to ask for your mother's forgiveness. She, in turn, asked for mine.

And maybe you saw the medal ceremony that was held for Jack Downey and Dick Fecteau at the Agency this past summer. It was moving, but the most moving part, for me, was the feeling that you were actually there beside me.

I hope you're proud of me, Joe. All those years as an Agency wife came in handy when I was trying to make it on my own as a party planner . . . cruise consultant . . . and floral boutique rep. (You can see I've had a lot of jobs.)

But nothing prepared me for life without you.

It's been twenty-three years since that day in the hospital when you started telling me your story—our story.

I kept my promise, Joe.

Bing